DEFENSE, SECURITY AND STRATEGIES

SECURITY OF UNITED STATES' NUCLEAR MATERIAL OVERSEAS

DEFENSE, SECURITY AND STRATEGIES

Additional books in this series can be found on Nova's website under the Series tab.

Additional e-books in this series can be found on Nova's website under the e-books tab.

DEFENSE, SECURITY AND STRATEGIES

SECURITY OF UNITED STATES' NUCLEAR MATERIAL OVERSEAS

RONALD E. ALLEN
AND
NANCY L. MILLER
EDITORS

NOVINKA
New York

For permission to use material from this book please contact us:
Telephone 631-231-7269; Fax 631-231-8175
Web Site: http://www.novapublishers.com

Additional color graphics may be available in the e-book version of this book.

LIBRARY OF CONGRESS CATALOGING-IN-PUBLICATION DATA

ISBN: 978-1-62417-038-6

Published by Nova Science Publishers, Inc. † New York

CONTENTS

PREFACE

The United States has exported special nuclear material, including enriched uranium, and source material such as natural uranium under nuclear cooperation agreements. The United States has 27 nuclear cooperation agreements for peaceful civilian cooperation. In order for the United States to engage in civilian nuclear cooperation with other states, it must conclude a framework agreement that meets specific requirements under Section 123 of the Atomic Energy Act (AEA). The AEA also provides for exemptions to these requirements, export control licensing procedures, and criteria for terminating cooperation. Congressional review is required for Section 123 agreements; the AEA establishes special parliamentary procedures by which Congress may act on a proposed agreement. This book discusses the security of United States' nuclear material overseas.

Chapter 1 – The United States has exported special nuclear material, including enriched uranium, and source material such as natural uranium under nuclear cooperation agreements. The United States has 27 nuclear cooperation agreements for peaceful civilian cooperation. Under the U.S. Atomic Energy Act of 1954 (AEA), as amended, partners are required to guarantee the physical protection of U.S. nuclear material. GAO was asked to (1) assess U.S. agency efforts to account for U.S. nuclear material overseas, (2) assess the Department of Energy's (DOE) and U.S. agencies' efforts to evaluate the security of U.S. material overseas, and (3) describe DOE's activities to secure or remove potentially vulnerable U.S. nuclear material at partner facilities. GAO analyzed agency records and interviewed DOE, Nuclear Regulatory Commission (NRC), Department of State (State), and partner country officials. This report summarizes GAO's classified report issued in June 2011.

Chapter 2 – In order for the United States to engage in civilian nuclear cooperation with other states, it must conclude a framework agreement that meets specific requirements under Section 123 of the Atomic Energy Act (AEA). The AEA also provides for exemptions to these requirements, export control licensing procedures, and criteria for terminating cooperation. Congressional review is required for Section 123 agreements; the AEA establishes special parliamentary procedures by which Congress may act on a proposed agreement.

In: Security of United States' Nuclear … ISBN: 978-1-62417-038-6
Editors: R. E. Allen and N. L. Miller

Chapter 1

NUCLEAR NONPROLIFERATION: U.S. AGENCIES HAVE LIMITED ABILITY TO ACCOUNT FOR, MONITOR, AND EVALUATE THE SECURITY OF U.S. NUCLEAR MATERIAL OVERSEAS*

The United States Government Accountability Office

ABBREVIATIONS

AEA	Atomic Energy Act
DOE	Department of Energy
EURATOM	European Atomic Energy Community
GTRI	Global Threat Reduction Initiative
HEU	highly enriched uranium
IAEA	International Atomic Energy Agency
IPPAS	International Physical Protection Advisory Service
LEU	low enriched uranium
NMMSS	Nuclear Materials Management and Safeguards System

* This is an edited, reformatted and augmented version of The United States Government Accountability Office publication, Report to the Committee on Foreign Affairs, House of Representatives GAO-11-920, dated September 2011.

NNPA	Nuclear Non-Proliferation Act of 1978
NNSA	National Nuclear Security Administration
NPT	Nuclear Nonproliferation Treaty
NRC	Nuclear Regulatory Commission
NSC	National Security Council
NSG	Nuclear Suppliers Group
SILEX	Separation of Isotopes by Laser Excitation
TRIGA	Training Research Isotope General Atomics

Why GAO Did This Study

The United States has exported special nuclear material, including enriched uranium, and source material such as natural uranium under nuclear cooperation agreements.

The United States has 27 nuclear cooperation agreements for peaceful civilian cooperation. Under the U.S. Atomic Energy Act of 1954 (AEA), as amended, partners are required to guarantee the physical protection of U.S. nuclear material. GAO was asked to (1) assess U.S. agency efforts to account for U.S. nuclear material overseas, (2) assess the Department of Energy's (DOE) and U.S. agencies' efforts to evaluate the security of U.S. material overseas, and (3) describe DOE's activities to secure or remove potentially vulnerable U.S. nuclear material at partner facilities. GAO analyzed agency records and interviewed DOE, Nuclear Regulatory Commission (NRC), Department of State (State), and partner country officials. This report summarizes GAO's classified report issued in June 2011.

What GAO Recommends

GAO suggests, among other things, that Congress consider directing DOE and NRC to compile an inventory of U.S. nuclear material overseas. DOE, NRC, and State generally disagreed with GAO's recommendations, including that they conduct annual inventory reconciliations with all partners, stating they were unnecessary.

GAO continues to believe that its recommendations could help improve the accountability of U.S. nuclear material in foreign countries.

What GAO Found

DOE, NRC, and State are not able to fully account for U.S. nuclear material overseas that is subject to nuclear cooperation agreement terms because the agreements do not stipulate systematic reporting of such information, and there is no U.S. policy to pursue or obtain such information. U.S. nuclear cooperation agreements generally require that partners report inventory information upon request, however, DOE and NRC have not systematically sought such data. DOE and NRC do not have a comprehensive, detailed, current inventory of U.S. nuclear material—including weapon-usable material such as highly enriched uranium (HEU) and separated plutonium—overseas that includes the country, facility, and quantity of material. In addition, NRC and DOE could not fully account for the current location and disposition of U.S. HEU overseas in response to a 1992 congressional mandate. U.S. agencies, in a 1993 report produced in response to the mandate, were able to verify the location of 1,160 kilograms out of 17,500 kilograms of U.S. HEU estimated to have been exported. DOE, NRC, and State have established annual inventory reconciliations with five U.S. partners, but not the others it has transferred material to or trades with.

Nuclear cooperation agreements do not contain specific access rights that enable DOE, NRC, or State to monitor and evaluate the physical security of U.S. nuclear material overseas, and the United States relies on its partners to maintain adequate security. In the absence of access rights, DOE's Office of Nonproliferation and International Security, NRC, and State have conducted physical protection visits to monitor and evaluate the physical security of U.S. nuclear material at facilities overseas when permitted. However, the agencies have not systematically visited countries believed to be holding the highest proliferation risk quantities of U.S. nuclear material, or systematically revisited facilities not meeting international physical security guidelines in a timely manner. Of the 55 visits made from 1994 through 2010, U.S. teams found that countries met international security guidelines approximately 50 percent of the time.

DOE has taken steps to improve security at a number of facilities overseas that hold U.S. nuclear material but faces constraints. DOE's Global Threat Reduction Initiative (GTRI) removes U.S. nuclear material from vulnerable facilities overseas but can only bring back materials that have an approved disposition pathway and meet the program's eligibility criteria. GTRI officials told GAO that, of the approximately 17,500 kilograms of HEU exported from the United States, 12,400 kilograms are currently not eligible for return to the United States. Specifically, GTRI reported that over 10,000 kilograms of U.S. HEU are believed to be in fuels from reactors in Germany, France, and Japan that have no disposition pathways in the United States and are adequately protected. In addition, according to GTRI, 2,000 kilograms of transferred U.S. HEU are located primarily in European Atomic Energy Community countries and are currently in use or adequately protected.

September 8, 2011

The Honorable Ileana Ros-Lehtinen
Chairman

The Honorable Howard L. Berman
Ranking Member
Committee on Foreign Affairs
House of Representatives

The United States has 27 nuclear cooperation agreements in force for peaceful civilian cooperation with partners including foreign countries, the European Atomic Energy Community (EURATOM), the International Atomic Energy Agency (IAEA), and Taiwan.[1] A nuclear cooperation agreement is a bilateral agreement that establishes the legal framework for significant civilian nuclear cooperation between the United States and other parties, including the transfer of certain nuclear material, including special nuclear material such as enriched uranium,[2] and source material such as natural uranium, nuclear reactors, and certain components of nuclear reactors.[3] The agreements are reciprocal, with both parties generally agreeing to all conditions specified in them. Under the U.S. Atomic Energy Act of 1954 (AEA), as amended, these agreements must contain certain obligations that govern, among other things, the U.S. rights of approval over the transfer, retransfer, enrichment, and reprocessing of certain kinds of nuclear material transferred from the United States and, in some cases produced overseas. A list of current and former U.S. nuclear cooperation agreement partners can be found in appendix II.

Thousands of kilograms of U.S. highly enriched uranium (HEU) and tens of thousands of kilograms of U.S. plutonium in spent fuel have accumulated overseas, as a result of foreign nuclear research and commercial nuclear power activities, which are subject to the terms in U.S. nuclear cooperation agreements. Inventories of U.S. nuclear material overseas could continue to grow as the result of reprocessing or recovery activities. As we have previously reported,[4] tracking and accounting for nuclear materials are important to (1) ensure that nuclear materials are used only for peaceful purposes, (2) comply with international treaty obligations, and (3) provide data to policymakers and other government officials. According to DOE and NRC officials, the United States obtains written assurances from partners in advance of each transfer of U.S.-obligated nuclear material that commits the partner to treat the transferred nuclear material according to the terms of its nuclear cooperation agreement with the United States. Nuclear material transferred from the United States, as well as special nuclear material produced overseas through the use of U.S.-supplied nuclear material or reactors, is known as "U.S.-obligated" material.

As the technology to design and create nuclear weapons has spread, one of the most serious threats facing the United States and other countries is the possibility that a nation or terrorist organization could steal weapon-usable nuclear materials from poorly secured stockpiles in various locations around the world.[5] In April 2009, President Obama made securing all vulnerable nuclear material worldwide within the next 4 years a key U.S. nonproliferation goal,[6] and the Department of Energy (DOE) has stated that doing so will require greater security cooperation with key countries; pursuing new partnerships to secure nuclear material; and strengthening nuclear security standards, practices, and international safeguards.

Securing nuclear material worldwide is a priority for DOE, through its National Nuclear Security Administration (NNSA),[7] which, as one of its core mission areas, aims to keep dangerous nuclear materials out of the hands of terrorists by securing nuclear weapons and nuclear materials at their source and by improving security practices around the world. In particular, two key DOE NNSA offices work with U.S. nuclear cooperation agreement partners to strengthen nuclear security practices and identify and secure vulnerable nuclear materials. First, DOE's Office of Nonproliferation and International Security works with countries to ensure that provisions in the agreements are met by, among other things, providing physical protection training, assessment, and guidance on a bilateral basis, and leading U.S. interagency physical protection visits to countries with U.S. nuclear material. Second, DOE's Office of Global Threat Reduction implements the Global Threat Reduction Initiative (GTRI) and, among other things, identifies, secures, removes, and/or facilitates the disposition of high-risk, vulnerable nuclear and radiological material at civilian sites around the world that pose a threat.

In November 2010, we reported on the export benefits facilitated by U.S. nuclear cooperation agreements.[8] In light of the quantities of nuclear material exported overseas under the framework of U.S. nuclear cooperation agreements, you asked us to also report on how U.S. agencies account for nuclear material overseas and monitor and evaluate the materials' physical security. Specifically, our objectives were to: (1) assess U.S. agency efforts to account for U.S. nuclear material overseas, (2) assess DOE's and other U.S. agencies' efforts to monitor and evaluate the physical security conditions of U.S. nuclear material subject to the terms of nuclear cooperation agreements, and (3) describe DOE's activities to secure or remove potentially vulnerable U.S. nuclear material at partner facilities. In June 2011, we reported to you on the results of our work in a classified report. This report summarizes certain aspects of our classified report.

To conduct this work, we reviewed relevant statutes, including the AEA, as amended, and the texts of all current U.S. nuclear cooperation agreements. We obtained data from the Nuclear Materials Management and Safeguards System (NMMSS), a database jointly operated by DOE and the Nuclear Regulatory Commission (NRC). This database, among other things, maintains data on U.S. peaceful use exports and retransfers of enriched uranium and plutonium that have occurred since 1950. To assess the reliability of data in NMMSS, we interviewed officials from DOE and NRC and a former DOE contractor to identify any limitations in NMMSS's data on the location and status of U.S. material overseas and found these data to be sufficiently reliable for the purposes of accounting for U.S. exports of nuclear material. We also compared NMMSS data with other sources of information regarding U.S. nuclear material transfers, including DOE data on nuclear material returns. We reviewed DOE, NRC, and Department of State (State) records and interviewed officials at those agencies to determine the extent to which DOE, NRC, and State are able to identify where U.S. nuclear material was exported, retransferred, and is currently held. We selected a non-probability sample of partners based on, among other considerations, quantities of U.S. special nuclear material transferred to them.[9] We conducted site visits in four countries that currently hold U.S.-obligated nuclear material and interviewed governmental officials and nuclear facility operators in these countries to discuss material accounting procedures and observe physical protection measures. Further, we interviewed officials from several other partners regarding their observations about working with the U.S. government to account for material subject to the terms of nuclear cooperation agreements. We reviewed IAEA's security guidelines, "The Physical Protection of Nuclear Material and Nuclear Facilities," INFCIRC/225/Rev.4,[10] and relevant international treaties. We also obtained and analyzed available records of U.S.

physical protection visits to partner facilities from 1974 through 2010. We reviewed agency documents and interviewed officials from DOE, NRC, and State to determine the process used for conducting physical protection visits at partner facilities. We obtained information from GTRI regarding its program's activities. We interviewed IAEA officials and reviewed relevant documents to better understand IAEA's role in maintaining safeguards and evaluating physical security measures. Additional details on our scope and methodology can be found in appendix I.

We conducted this performance audit from September 2010 to June 2011 in accordance with generally accepted government auditing standards. Those standards require that we plan and perform the audit to obtain sufficient, appropriate evidence to provide a reasonable basis for our findings and conclusions based on our audit objectives. We believe that the evidence obtained provides a reasonable basis for our findings and conclusions based on our audit objectives.

BACKGROUND

The AEA, as amended, sets forth the procedures and requirements for the U.S. government's negotiating, proposing, and entering into nuclear cooperation agreements with foreign partners. The AEA, as amended, requires that U.S. peaceful nuclear cooperation agreements contain the following nine provisions: [11]

1. *Safeguards:* Safeguards, as agreed to by the parties, are to be maintained over all nuclear material and equipment transferred, and all special nuclear material used in or produced through the use of such nuclear material and equipment, as long as the material or equipment remains under the jurisdiction or control of the cooperating party, irrespective of the duration of other provisions in the agreement or whether the agreement is terminated or suspended for any reason. Such safeguards are known as "safeguards in perpetuity."
2. *Full-scope IAEA safeguards as a condition of supply:* In the case of non-nuclear weapons states, continued U.S. nuclear supply is to be conditioned on the maintenance of IAEA "full-scope" safeguards over all nuclear materials in all peaceful nuclear activities within the territory, under the jurisdiction, or subject to the control of the cooperating party.[12]

3. *Peaceful use guaranty:* The cooperating party must guarantee that it will not use the transferred nuclear materials, equipment, or sensitive nuclear technology, or any special nuclear material produced through the use of such, for any nuclear explosive device, for research on or development of any nuclear explosive device, or for any other military purpose.
4. *Right to require return:* An agreement with a non-nuclear weapon state must stipulate that the United States has the right to require the return of any transferred nuclear materials and equipment, and any special nuclear material produced through the use thereof, if the cooperating party detonates a nuclear device, or terminates or abrogates an agreement providing for IAEA safeguards.
5. *Physical security*: The cooperating party must guarantee that it will maintain adequate physical security for transferred nuclear material and any special nuclear material used in or produced through the use of any material, or production or utilization facilities transferred pursuant to the agreement.[13]
6. *Retransfer rights:* The cooperating party must guarantee that it will not transfer any material, Restricted Data, or any production or utilization facility transferred pursuant to the agreement, or any special nuclear material subsequently produced through the use of any such transferred material, or facilities, to unauthorized persons or beyond its jurisdiction or control, without the consent of the United States.
7. *Restrictions on enrichment or reprocessing of U.S.-obligated material:* The cooperating party must guarantee that no material transferred, or used in, or produced through the use of transferred material or production or utilization facilities, will be reprocessed or enriched, or with respect to plutonium, uranium-233, HEU, or irradiated nuclear materials, otherwise altered in form or content without the prior approval of the United States.
8. *Storage facility approval:* The cooperating party must guarantee not to store any plutonium, uranium-233, or HEU that was transferred pursuant to a cooperation agreement, or recovered from any source or special nuclear material transferred, or from any source or special nuclear material used in a production facility or utilization facility transferred pursuant to the cooperation agreement, in a facility that has not been approved in advance by the United States.

9. *Additional restrictions:* The cooperating party must guarantee that any special nuclear material, production facility, or utilization facility produced or constructed under the jurisdiction of the cooperating party by or through the use of transferred sensitive nuclear technology, will be subject to all the requirements listed above.

In addition, the United States is a party to the Treaty on the Non-Proliferation of Nuclear Weapons (NPT). The NPT binds each of the treaty's signatory states that had not manufactured and exploded a nuclear weapon or other nuclear explosive device prior to January 1, 1967 (referred to as non-nuclear weapon states) to accept safeguards as set forth in an agreement to be concluded with IAEA. Under the safeguards system, IAEA, among other things, inspects facilities and locations containing nuclear material, as declared by each country, to verify its peaceful use. IAEA standards for safeguards agreements provide that the agreements should commit parties to establish and maintain a system of accounting for nuclear material, with a view to preventing diversion of nuclear energy from peaceful uses, and reporting certain data to IAEA.

IAEA's security guidelines provide the basis by which the United States and other countries generally classify the categories of protection that should be afforded nuclear material, based on the type, quantity, and enrichment of the nuclear material. For example, Category I material is defined as 2 kilograms or more of unirradiated or "separated" plutonium or 5 kilograms of uranium-235 contained in unirradiated or "fresh" HEU and has the most stringent set of recommended physical protection measures. The recommended physical protection measures for Category II and Category III nuclear materials are less stringent. Appendix III contains further details on the categorization of nuclear material.

DOE, NRC, and State Are Not Able to Fully Account for U.S. Nuclear Material Located at Foreign Facilities

DOE, NRC, and State are not able to fully account for U.S. nuclear material overseas that is subject to nuclear cooperation agreement terms because the agreements do not stipulate systematic reporting of such information, and there is no U.S. policy to pursue or obtain such information.

Section 123 of the AEA, as amended, does not require nuclear cooperation agreements to contain provisions stipulating that partners report information on the amount, status, or location (facility) of special nuclear material subject to the agreement terms. However, U.S. nuclear cooperation agreements generally require that partners report inventory information upon request, although DOE and NRC have not systematically sought such data. We requested from multiple offices at DOE and NRC a current and comprehensive inventory of U.S. nuclear material overseas, to include country, site, or facility, and whether the quantity of material was rated as Category I or Category II material. However, neither agency has provided such an inventory. NMMSS does not contain the data necessary to maintain an inventory of U.S. special nuclear material overseas. DOE, NRC, and State have not pursued annual inventory reconciliations of nuclear material subject to U.S. cooperation agreement terms with all foreign partners that would provide the U.S. government with better information about where such material is held. Furthermore, according to DOE, NRC, and State officials, no U.S. law or policy directs U.S. agencies to obtain information regarding the location and disposition of U.S. nuclear material at foreign facilities.

U.S. Nuclear Cooperation Agreements Generally Require That Partners Report Inventory Information upon Request, but DOE and NRC Have Not Systematically Sought Such Data

Section 123 of the AEA, as amended, does not require nuclear cooperation agreements to contain provisions stipulating that partners report information on the amount, status, or location (facility) of special nuclear material subject to the agreement terms. However, the texts of most U.S. nuclear cooperation agreements contain a provision calling for each partner to maintain a system of material accounting and control and to do so consistent with IAEA safeguards standards or agreements.[14] In addition, we found that all agreements, except three negotiated prior to 1978 and the U.S.-China agreement, contain a provision that the other party shall report, or shall authorize the IAEA to report, inventory information upon request. However, according to DOE and NRC officials, with the exception of the administrative arrangements with five partners, the United States has not requested such information from all partners on an annual or systematic basis.

Nonetheless, the AEA requires U.S. nuclear cooperation agreements to include terms that, among other things, obligate partners to obtain U.S.

approval for the transfer, retransfer, enrichment and reprocessing, and the storage of U.S.-obligated uranium-233, HEU, or other nuclear materials that have been irradiated. In addition, according to DOE and NRC officials, the United States obtains written assurances from partners in advance of each transfer of U.S. nuclear material that commits them to maintain the transferred nuclear material according to the terms of its nuclear cooperation agreement with the United States. DOE and NRC officials told us these assurances help the United States ensure that partner countries comply with the terms of the nuclear cooperation agreement.

In addition, IAEA, DOE, NRC, and State officials told us that IAEA's safeguards activities provide a level of assurance that nuclear material is accounted for at partner facilities. The safeguards system, which has been a cornerstone of U.S. efforts to prevent nuclear proliferation, allows IAEA to independently verify that non-nuclear weapons states that signed the NPT are complying with its requirements. Under the safeguards system, IAEA, among other things, inspects facilities and locations containing nuclear material declared by countries to verify its peaceful use. Inspectors from IAEA's Department of Safeguards verify that the quantities of nuclear material that these non-nuclear weapons states declared to IAEA are not diverted for other uses. IAEA considers such information confidential and does not share it with its member states, including the United States, unless the parties have agreed that IAEA can share the information.

IAEA's inspectors do not verify nuclear material by country of origin or associated obligation. DOE, State, and IAEA officials told us that, because IAEA does not track the obligation of the material under safeguards, IAEA may notice discrepancies in nuclear material balances through periodic reviews of countries' shipping records. However, these officials said that IAEA does not have the ability to identify whether and what volume of nuclear material at partner country facilities is U.S.- obligated and therefore subject to the terms of U.S. nuclear cooperation agreements.

DOE and NRC Do Not Have a Current Comprehensive Inventory of U.S. Material Overseas

DOE and NRC do not have a comprehensive, detailed, current inventory of U.S. nuclear material overseas that would enable the United States to identify material subject to U.S. nuclear cooperation agreement terms. We requested from multiple offices at DOE and NRC a current and comprehensive inventory of U.S. nuclear material overseas, to include country, site, or facility, and whether the quantity of material was Category I or Category II. However, the agencies have not provided such a list. DOE officials from the Office of Nonproliferation and International Security told us that they have multiple mechanisms to account for the amount of U.S.-obligated nuclear material at foreign facilities. They stated that they use NMMSS records to obtain information regarding U.S. nuclear material inventories held in other countries. However, NMMSS officials told us that NMMSS was an accurate record of material exports from the United States, but that it should not be used to estimate current inventories. In addition, NMMSS officials stated that DOE's GTRI program has good data regarding the location of U.S. nuclear material overseas and that this information should be reconciled with NMMSS data.

However, when we requested information regarding the amount of U.S. material at partner facilities, GTRI stated that they could not report on the amount of U.S. nuclear material remaining at facilities unless it was scheduled for GTRI to return. In addition, in February 2011 written comments to us, GTRI stated it was not responsible for acquiring or maintaining inventory information regarding U.S. nuclear material overseas. A long-time contract employee for DOE's Office of Nonproliferation and International Security stated he has tried to collect information regarding U.S. nuclear material overseas from various sources including a list of countries eligible for GTRI's fuel return program, NMMSS, and other sources, but it is not possible to reconcile information from the various lists and sources and consequently there is no list of U.S. inventories overseas.

According to public information, the United States has additional measures known as administrative arrangements with five of its trading partners to conduct annual reconciliations of nuclear material amounts. In addition, for all partners, DOE and NRC officials told us that an exchange of diplomatic notes is sent prior to any transfer to ensure that U.S. nuclear material is not diverted for non-peaceful purposes, and which binds the partner to comply with the terms of the nuclear cooperation agreement. However, the measures cited by DOE are not comprehensive or sufficiently detailed to provide the specific location of U.S. nuclear material overseas.

NRC and DOE could not fully account for U.S. exports of HEU in response to a congressional mandate that the agencies report on the current location and disposition of U.S. HEU overseas. In 1992, Congress mandated that NRC,[15] in consultation with other relevant agencies, submit to Congress a report detailing the current status of previous U.S. exports of HEU, including its location, disposition (status), and how it had been used. The January 1993 report that NRC produced in response to the mandate stated it was not possible to reconcile this information from available U.S. sources of data with all foreign holders of U.S. HEU within the 90-day period specified in the act.[16] The report further states that a thorough reconciliation of U.S and foreign records with respect to end use could require several months of additional effort, assuming that EURATOM would agree to participate. According to DOE and NRC officials, no further update to the report was issued, and the U.S. government has not subsequently attempted to develop such a comprehensive estimate of the location and status of U.S. HEU overseas.

The 1993 report provided estimated material balances based on the transfer, receipt, or other adjustments reported to the NMMSS and other U.S. agencies. The report stated that the estimated material balances should match partners' reported inventories. However, the report did not compare the balances or explain the differences.

Our analysis of other documentation associated with the report shows that NRC, in consultation with U.S. agencies, was able to verify the location of 1,160 kilograms out of an estimated 17,500 kilograms of U.S. HEU remaining overseas as of January 1993. NRC's estimates matched partner estimates in 22 cases; did not match partner estimates in 6 cases; and, in 8 cases, partners did not respond in time to NRC's request.

The 1993 report noted that, in cases where U.S. estimates did not match partners' inventory reports, "reconciliation efforts are underway." However, DOE, NRC, and NMMSS officials told us that no further report was issued. In addition, NMMSS officials told us that they were unaware of any subsequent efforts to reconcile U.S. estimates with partners' reports, or update the January 1993 report. In addition, we found no indication that DOE, NMMSS, or NRC officials have updated the January 1993 report, or undertaken a comprehensive accounting of U.S. nuclear material overseas.

NMMSS Does Not Contain Data Necessary to Identify Where U.S. Material Is Located Overseas

We found that NMMSS does not contain the data necessary to maintain an inventory of U.S. nuclear material overseas subject to U.S. nuclear cooperation agreements. According to NRC documents, NMMSS is part of an overall program to help satisfy the United States' accounting, controlling, and reporting obligations to IAEA and its nuclear trading partners. NMMSS, the official central repository of information on domestic inventories and exports of U.S. nuclear material, contains current and historic data on the possession, use, and shipment of nuclear material. It includes data on U.S.-supplied nuclear material transactions with other countries and international organizations, foreign contracts, import/export licenses, government-to-government approvals, and other DOE authorizations such as authorizations to retransfer U.S. nuclear material between foreign countries.[17] DOE and NRC officials told us that NMMSS contains the best available information regarding U.S. exports and retransfers of special nuclear material.

DOE and NRC do not collect data necessary for NMMSS to keep an accurate inventory of U.S. nuclear material overseas. According to NRC officials, NMMSS cannot track U.S. nuclear material overseas because data regarding the current location and status of U.S. nuclear material, such as irradiation, decay, burn up, or production, are not collected. NMMSS only contains data on domestic inventories and transaction receipts from imports and exports reported by domestic nuclear facilities and some retransfers reported by partners to the United States and added to the system by DOE. Therefore, while the 1995 Nuclear Proliferation Assessment Statement accompanying the U.S.-EURATOM agreement estimated 250 tons of U.S.-obligated plutonium are planned to be separated from spent power reactor fuel in Europe and Japan for use in civilian energy programs in the next 10 to 20 years, our review indicates that the United States would not be able to identify the European countries or facilities where such U.S.-obligated material is located.

DOE, NRC, and State Have Not Pursued Annual Reconciliations of Inventories of Nuclear Material Subject to U.S. Nuclear Cooperation Agreement Terms with All Partners

DOE, NRC, and State have not pursued annual inventory reconciliations of nuclear material subject to U.S. nuclear cooperation agreement terms with all partners that would provide the U.S. government with better information about where such material is held overseas. Specifically, once a nuclear cooperation agreement is concluded, U.S. government officials—generally led by DOE—and partner country officials may negotiate an administrative arrangement for an annual inventory reconciliation to exchange information regarding each country's nuclear material accounting balances. Inventory reconciliations typically compare the countries' data and material transfer and retransfer records, and can help account for material consumed or irradiated by reactors.

Government officials from several leading nuclear material exporting and importing countries told us that they have negotiated with all their other partners to exchange annual inventory reconciliations to provide a common understanding of the amount of their special material held by another country or within their country. For example, Australia, which exports about 13 percent of the world's uranium each year, conducts annual reconciliations with each of its partners, and reports annually to the Australian Parliament regarding the location and disposition of all Australian nuclear material. NRC officials told us that Australia has some of the strictest reporting requirements for its nuclear material.

The United States conducts annual inventory reconciliations with five partners but does not conduct inventory reconciliations with the other partners it has transferred material to or trades with.[18] According to DOE officials, for the five reconciliations currently conducted, NMMSS data are compared with the partner's records and, if warranted, each country's records are adjusted, where necessary, to reflect the current status of U.S special nuclear material. As of February 2011, the United States conducted bilateral annual exchanges of total material balances for special nuclear materials with five partners. Of these partners, the United States exchanges detailed information regarding inventories at each specific facility only with one partner. DOE officials noted that they exchange information with particular trading partners on a transactional basis during the reporting year and work with the partners at that time to resolve any potential discrepancies that may arise. In the case of EURATOM, material information is reported as the cumulative total of all 27 EURATOM members. For the purposes of nuclear cooperation with the United States, EURATOM is treated as one entity rather than its 27 constituent parts. None of the 27 EURATOM member states have bilateral nuclear cooperation agreements in force with the United States.

According to a 2010 DOE presentation for NMMSS users, the difference in reporting requirements results in a 69-page report for Japan and a 1-page report for EURATOM. In addition, information exchanged with other trading partners also is not reported by facility. DOE and NRC officials told us that the United States may not have accurate information regarding the inventories of U.S. nuclear material held by its 21 other partners.

DOE officials told us that, in addition to benefits, there were costs to pursuing facility-by-facility reconciliations and reporting. In particular, DOE officials told us they have not pursued facility-by-facility accounting in annual reconciliations with other partners because it would be difficult for the United States to supply such detailed information regarding partner material held in U.S. facilities. DOE and NRC officials told us this would also create an administrative burden for the United States. According to DOE officials, the relative burden with which the United States can perform facility-by-facility accounting by foreign trading partner varies greatly based on the amount of material in the United States that is obligated to such partners. For example, the United States can perform facility-by-facility accounting with one country, because U.S. officials told us there is not much of that country's nuclear material in the United States. However, if the United States were to conduct facility-by-facility accounting with Australia, it would create burdensome reporting requirements. Specifically, according to DOE officials, Australia would have to report to the United States on the status of a few facilities holding U.S. nuclear material, but the United States would be required to report on hundreds of U.S. facilities holding Australian nuclear material. Without information on foreign facilities, however, it may be difficult to track U.S. nuclear materials for accounting and control purposes.

No U.S. Law or Policy Directs U.S. Agencies to Obtain Information Regarding the Location and Disposition of U.S. Nuclear Material at Foreign Facilities

DOE, NRC, and State officials told us neither U.S. law nor U.S. policy explicitly requires the United States to track U.S. special nuclear material overseas. Moreover, U.S. law does not require peaceful nuclear cooperation agreements to require cooperating parties to provide reports to the United States of nuclear material on a facility-by-facility basis. A March 2002 DOE Inspector General's audit raised concerns about the U.S. government's ability to track sealed sources, which could contain nuclear or radioactive material.[19] In response to the audit's findings, NNSA's Associate Administrator for Management and Administration wrote that "While it is a good idea to be aware of the locations and conditions of any [U.S.] material, it is not the current policy of the U.S. government." Furthermore, the Associate Administrator asserted that various U.S. government agencies, including State, DOE, and NRC, would need to be involved should DOE change its policy and undertake an initiative to track the location and condition of U.S. sealed sources in foreign countries. Similarly, DOE, NRC, and State officials told us that if it became the policy of the U.S. government to track nuclear material overseas—and in particular, by facility—then requirements would have to be negotiated into the nuclear cooperation agreements or the associated administrative arrangements.

NMMSS officials told us that NMMSS is currently capable of maintaining information regarding inventories of U.S. nuclear material overseas. However, as we reported in 1982,[20] NMMSS information is not designed to track the location (facility) or the status—such as whether the material is irradiated or unirradiated, fabricated into fuel, burned up, or reprocessed.

As a result, NMMSS neither identifies where U.S. material is located overseas nor maintains a comprehensive inventory of U.S.- obligated material. In addition, NMMSS officials emphasized that this information would need to be systematically reported.

According to these officials, such reporting is not done on a regular basis by other DOE offices and State. In some instances, State receives a written notice of a material transfer at its embassies and then transmits this notice to DOE.

Officials from DOE's Office of Nonproliferation and International Security told us that, while they could attempt to account for U.S. material overseas on a case-by-case basis, obtaining the information to systematically track this material would require renegotiating the terms of nuclear cooperation agreements.

DOE has recently issued proposed guidance clarifying the role of DOE offices for maintaining and controlling U.S. nuclear material. An October 2010 draft DOE order states that DOE "Manages the development and maintenance of NMMSS by: (a) collecting data relative to nuclear materials including those for which the United States has a safeguards interest both domestically and abroad; (b) processing the data; and (c) issuing reports to support the safeguards and management needs of DOE and NRC, and other government organizations, including those associated with international treaties and organizations."[21]

However, we did not find any evidence that DOE will be able to meet those responsibilities in the current configuration of NMMSS without obtaining additional information from partners and additional and systematic data sharing among DOE offices.

DOE, NRC, AND STATE DO NOT HAVE ACCESS RIGHTS TO MONITOR AND EVALUATE THAT U.S. NUCLEAR MATERIAL LOCATED AT FOREIGN FACILITIES IS ADEQUATELY PROTECTED

Nuclear cooperation agreements do not contain specific access rights that enable DOE, NRC, or State to monitor and evaluate the physical security of U.S. nuclear material overseas, and the United States relies on partners to maintain adequate security. In the absence of specific access rights, DOE, NRC, and State have jointly conducted interagency physical protection visits to monitor and evaluate the physical security of nuclear material when given permission by the partner country. However, the interagency physical protection teams have neither systematically visited countries believed to be holding Category I quantities of U.S. nuclear material, nor have they systematically revisited facilities determined to not be meeting IAEA security guidelines in a timely manner.

U.S. Agencies' Ability to Evaluate the Security of U.S. Nuclear Material Overseas Is Limited by Lack of Access Rights, and the United States Relies on Partners to Maintain Adequate Security

DOE's, NRC's, and State's ability to monitor and evaluate whether material subject to U.S. nuclear cooperation agreement terms is physically secure is contingent on partners granting access to facilities where such material is stored. Countries, including the United States, believe that the physical protection of nuclear materials is a national responsibility. This principle is reflected both in IAEA's guidelines on the "Physical Protection of Nuclear Material and Nuclear Facilities" and in pending amendments to the Convention on the Physical Protection of Nuclear Material. Our review of section 123 of the AEA and all U.S. nuclear cooperation agreements currently in force found that they do not explicitly include a provision granting the United States access to verify the physical protection of facilities or sites holding material subject to U.S. nuclear cooperation agreement terms. However, in accordance with the AEA, as amended, all nuclear cooperation agreements, excepting three negotiated prior to 1978, contain provisions requiring both partners to maintain adequate physical security over transferred material.

The AEA, as amended, requires that the cooperating party must guarantee that it will maintain adequate physical security for transferred nuclear material and any special nuclear material used in or produced through the use of any material, or production, or utilization facility transferred pursuant to the agreement. However, it does not specify that State, in cooperation with other U.S. agencies, negotiates agreement terms that must include rights of access or other measures for the United States to verify whether a partner is maintaining adequate physical security over U.S. material. Our review of the texts of all 27 U.S. nuclear cooperation agreements in force found that most of them contain a provision providing that the adequacy of physical protection measures shall be subject to review and consultations by the parties. However, none of the agreements include specific provisions stipulating that the United States has the right to verify whether a partner is adequately securing U.S. nuclear material.[22] As a result, several DOE and State officials told us the United States' ability to monitor and evaluate the physical security of U.S. nuclear material overseas is contingent on partners' cooperation and access to facilities where U.S. material is stored.

State, DOE, and NRC officials told us that they rely on partners to comply with IAEA's security guidelines for physical protection. However, the guidelines, which are voluntary, do not provide for access rights for other states to verify whether physical protection measures for nuclear material are adequate. IAEA's security guideline document states that the "responsibility for establishing and operating a comprehensive physical protection system for nuclear materials and facilities within a State rests entirely with the Government of that State." In addition, according to the guidelines, member states should ensure that their national laws provide for the proper implementation of physical protection and verify continued compliance with physical protection regulations. For example, according to IAEA's security guidelines, a comprehensive physical protection system to secure nuclear material should include, among other things,

- technical measures such as vaults, perimeter barriers, intrusion sensors, and alarms;
- material control procedures; and
- adequately equipped and appropriately trained guard and emergency response forces.

In addition, according to DOE and State officials, key international treaties, including the Convention on the Physical Protection of Nuclear

Material—which calls for signatory states to provide adequate physical protection of nuclear material while in international transit—do not provide states the right to verify the adequacy of physical protection measures. A senior official from IAEA's Office of Nuclear Security told us that physical security is a national responsibility and that governments may choose to organize their various physical security components differently, as long as the components add up to an effective regime.

Despite these constraints on access, the U.S. government can take certain actions to protect U.S. nuclear material located at foreign facilities. For example, NRC licensing for the export of nuclear equipment and material is conditioned on partner maintenance of adequate physical security. NRC officials stated that, when an export license application for nuclear materials or equipment is submitted, the U.S. government seeks confirmation, in the form of peaceful use assurances, from the foreign government that the material and equipment, if exported, will be subject to the terms and conditions of that government's nuclear cooperation agreement with the United States. In addition, NRC officials stated that this government-to-government reconfirmation of the terms and conditions of the agreement meets the "letter and spirit" of the AEA and Nuclear Non-Proliferation Act of 1978 (NNPA) and underscores that the partner is aware of and accepts the terms and conditions of the agreement.

NRC officials also noted that the NNPA amendments to the AEA were designed and intended to encourage foreign governments to agree to U.S. nonproliferation criteria in exchange for nuclear commodities. However, the AEA does not empower the U.S. government through inspections or other means to enforce foreign government compliance with nuclear cooperation agreements once U.S. nuclear commodities are in a foreign country. Importantly, according to NRC, the onus is on the receiving country as a sovereign right and responsibility and consistent with its national laws and international commitments, to adequately secure the nuclear material.

According to DOE and State, as well as foreign government officials, the United States and the partner share a strong common interest in deterring and preventing the misuse of nuclear material, as well as an interest in maintaining the rights afforded to sovereign countries. The partner's interest in applying adequate security measures, for instance, is particularly strong because the nuclear material is located within its territory. Moreover, specific physical security needs may often depend on unique circumstances and sensitive intelligence information known only to the partner.

In addition, the AEA requires that U.S. nuclear cooperation agreements with non-nuclear weapon states contain a stipulation that the United States shall have the right to require the return of certain nuclear material, as well as equipment, should the partner detonate a nuclear device or terminate or abrogate its safeguards agreements with IAEA. However, DOE, NRC, and State officials told us that the U.S. government has never exercised the "right to require return" provisions in its nuclear cooperation agreements. In addition, the United States typically includes "fall-back safeguards"—contingency plans for the application of alternative safeguards should IAEA safeguards become inapplicable for any other reason. DOE and State officials told us, however, that the United States has not exercised its fall-back safeguards provisions, because the United States has not identified a situation where IAEA was unable to perform its safeguards duties.

U.S. Agencies Have Visited Foreign Sites to Monitor and Evaluate the Physical Security of U.S. Nuclear Material

U.S. agencies have, over time, made arrangements with partners to visit certain facilities where U.S. nuclear material is stored. As we reported in August 1982 and in December 1994,[23] U.S. interagency physical protection teams visit partner country facilities to monitor and evaluate whether the physical protection provided to U.S. nuclear material meets IAEA physical security guidelines. In 1974, DOE's predecessor, the Energy Research and Development Administration, began leading teams composed of State, NRC, and DOE national laboratory officials to review the partner's legal and regulatory basis for physical protection and to ensure that U.S. nuclear material was adequately protected. In 1988, the Department of Defense's Defense Threat Reduction Agency began to participate in these visits, and officials from other agencies and offices, such as GTRI, have participated. The visits have generally focused on research reactors containing HEU but have also included assessments, when partners voluntarily grant access, of other facilities' physical security, including nuclear power plants, reprocessing facilities, and research and development facilities containing U.S. nuclear material.

According to DOE documents and DOE, NRC, and State officials, the primary factors for selecting countries for visits are the type, quantity, and form of nuclear material, with priority given to countries with U.S. HEU or plutonium in Category I amounts. In addition, in 1987, NRC recommended that countries possessing U.S. Category I nuclear material be revisited at least every 5 years. DOE and NRC officials told us this has become an official goal for prioritizing visits. According to DOE, interagency physical protection visits are also made whenever the country has had or expects to have a significant change in its U.S. nuclear material inventory, along with other factors, such as previous findings that physical protection was not adequate. These criteria and other factors are used to help U.S. agencies prioritize visits on a countrywide basis and also supplement other information that is known about a partner's physical protection system and the current threat environment. Moreover, while the U.S. physical protection program assesses physical security conditions on a site-specific basis, NRC's regulations permit the determination of adequacy of foreign physical protection systems on a countrywide basis.[24] Therefore, DOE, NRC, and State officials told us that the results of the interagency physical protection visits, combined with other sources of information such as country threat assessments, are used as a measure of the physical security system countrywide.

The U.S. teams visit certain facilities where U.S. nuclear material is used or stored to observe physical protection measures after discussing the relevant nuclear security regulatory framework with the partner government. DOE and State officials told us these physical protection visits help U.S. officials develop relationships with partner officials, share best practices and, in some cases, recommend physical security improvements.

We visited four facilities that hold U.S.-obligated nuclear material. The partner officials and facility operators we met shared their observations regarding the U.S. physical protection visits. Representatives from one site characterized a recent interagency physical protection visit as a "tour." These officials told us the U.S. government officials had shared some high-level observations regarding their visit with government officials and nuclear reactor site operators but did not provide the government or site operators with written observations or recommendations. On the other hand, government officials from another country we visited told us that a recent interagency physical protection visit had resulted in a useful and detailed exchange of information about physical security procedures. These government officials told us they had learned "quite a lot" from the interagency physical protection visit and that they hoped the dialogue would continue, since security could always be improved. In February 2011, DOE officials told us they had begun to distribute the briefing slides they use at the conclusion of a physical protection visit to foreign officials. State officials told us that the briefings are considered government-to-government activities, and it is the partner government's choice on whether to include facility operators in the briefings.

In addition, we reviewed U.S. agencies' records of these and other physical protection visits and found that, over the 17-year period from 1994 through 2010, U.S. interagency physical protection teams made 55 visits. Of the 55 visits, interagency physical protection teams found the sites met IAEA security guidelines on 27 visits, did not meet IAEA security guidelines on 21 visits, and the results of 7 visits are unknown because the physical protection team was unable to assess the sites, or agency documentation was missing.

According to DOE, State, and NRC officials, the visits are used to encourage security improvements by the partner. For example, based on the circumstances of one particular facility visited in the last 5 years, the physical protection team made several recommendations to improve security, including installing (1) fences around the site's perimeter, (2) sensors between fences, (3) video assessment systems for those sensors, and (4) vehicle barriers. According to DOE officials, these observations were taken seriously by the country, which subsequently made the improvements.

When we visited the site as part of our review, government officials from that country told us the U.S. interagency team had provided useful advice and, as a result, the government had approved a new physical protection plan. These government officials characterized their interactions with DOE and other U.S. agency officials as positive and told us that the government's new physical protection plan had been partly implemented. Moreover, although we were not granted access to the building, we observed several physical protection upgrades already implemented or in progress, including: (1) the stationing of an armed guard outside the facility holding U.S. Category I material; (2) ongoing construction of a 12- foot perimeter fence around the facility; and (3) construction of a fence equipped with barbed wire and motion detectors around the entire research complex. We were also told that, among other things, remote monitoring equipment had been installed in key areas in response to the interagency visit. The Central Alarm Station was hardened, and the entrance to the complex was controlled by turnstiles and a specially issued badge, which entrants received after supplying a passport or other government-issued identification. Private automobiles were not allowed in the facility.

Not all U.S. physical protection visits proceed smoothly. In some cases, U.S. agencies have attempted repeatedly to convince partner officials of the seriousness of meeting IAEA security guidelines and to fund improvements. For example, a U.S. interagency physical protection team in the early 2000s found numerous security problems at a certain country's research reactor. The site supervisor objected to the interagency team's assessment because physical security was a matter of national sovereignty, and IAEA security guidelines were subject to interpretation. The site supervisor also objected to some of the U.S. team's recommendations. In some instances, under U.S. pressure, countries have agreed to make necessary improvements with DOE technical and material assistance.

Our review of agency records indicates that, in recent years, as the number of countries relying on U.S. HEU to fuel research reactors has continued to decline, U.S. agencies have succeeded in using a partner's pending export license for U.S. HEU or expected change in inventory of U.S. special nuclear material as leverage for a U.S. interagency physical protection visit. For example, we identified two cases since 2000 where a partner country applied for a license to transfer U.S. HEU, and a U.S. interagency team subsequently visited those two sites. In addition, we identified a recent situation where a partner country's inventory of U.S. plutonium at a certain site was expected to significantly increase, and a U.S. interagency team visited the site to determine whether the site could adequately protect these additional inventories.

According to DOE officials, requests for U.S. low enriched uranium (LEU) export licenses have increased in recent years.[25] In response, DOE officials told us that U.S. agencies have begun to prioritize visits to countries making such requests, and our review of agency documentation corroborates this. For example, physical protection visit records we reviewed state that recent interagency physical protection visits were made to two sites to evaluate the facilities' physical security in advance of pending U.S. LEU license applications. In addition, a DOE contractor and State official told us that a U.S. team planned to visit another partner country site in late 2011 in order to verify the adequacy of physical protection for U.S.-obligated LEU.

U.S. Agencies Do Not Have a Formal Process for Coordinating and Prioritizing U.S. Physical Protection Visits

DOE, NRC, and State do not have a formal process for coordinating and prioritizing U.S. interagency physical protection visits. In particular, DOE, which has the technical lead and is the agency lead on most visits has neither (1) worked with NRC and State to establish a plan and prioritize interagency physical protection visits, nor (2) measured performance in a systematic way. Specifically:

- *Establishing a plan and prioritizing and coordinating efforts.* A U.S. agency formal plan for which countries or facilities to visit has not been established, nor have goals for the monitoring and evaluation activities been formalized. In October 2009, DOE reported to us that it had formulated a list of countries that contained U.S. nuclear material and were priorities for U.S. teams to visit. However, in a subsequent written communication to us, a senior DOE official stated that DOE had not yet discussed this list with State, NRC, or other agency officials. [26] As a result, the list of countries had not been properly vetted at that time and did not represent an interagency agreed upon list. In February 2011, DOE officials told us that U.S. agencies will be considering a revised methodology for prioritizing physical protection visits. NRC officials told us they thought the interagency coordination and prioritization of the visit process could be improved. A State official, who regularly participates in the U.S. physical protection visits, told us that interagency coordination had improved in the past 6 months, in response to a recognized need by U.S. agencies to be prepared for an expected increase in requests for exports of U.S. LEU.
- *Measuring performance.* The agencies have not developed performance metrics to gauge progress in achieving stated goals related to physical protection visits. Specifically, DOE, NRC, and State have not performed an analysis to determine whether the stated interagency goal of visiting countries containing U.S. Category I nuclear material within 5 years has been met. In addition, although DOE has stated U.S. physical protection teams revisit sites whenever there is an indication that security does not meet IAEA security guidelines, DOE has not quantified its efforts in a meaningful way. In response to our questions about metrics, DOE officials stated that there is no U.S. law regarding the frequency of visits or revisits and that the agency's internal goals are not requirements. These officials told us that DOE, NRC, and State recognize that the "number one goal" is to ensure the physical security of U.S. nuclear material abroad. DOE officials stated that the best measure of the U.S. physical protection visits' effectiveness is that there has not been a theft of U.S. nuclear material from a foreign facility since the 1970s, when two LEU fuel rods were stolen from a certain country. However, officials reported to us that, in 1990, the facility was determined to be well below IAEA security guidelines. Our review of DOE documentation shows that other U.S. LEU transferred to the facility remains at the site.

In July 2011, in conjunction with the classification review for this report, DOE officials stated that while DOE, NRC, and State work together on coordinating U.S. government positions regarding priorities and procedures for the interagency physical protection program, no updated document exists that formalizes the process for planning, coordinating, and prioritizing U.S. interagency physical protection visits. We note that the documents that DOE refers to are internal DOE documents presented to us in 2008 and 2009 in response to questions regarding nuclear cooperation agreements. These documents are not an interagency agreed-upon document, but reflect DOE's views on determining which countries and facilities interagency physical protection teams should visit. Further, DOE officials in July 2011 stated that DOE, NRC, and State do not have an agreed-upon way to measure performance in a systematic way, and that while the goals for the monitoring and evaluation activities have not yet been formalized through necessary updated documents, a prioritized list of countries to visit does exist. These officials noted that the U.S. government is working to update its planning documents and is examining its methodology for prioritizing physical protection visits. Any changes will be included in these updated documents.

DOE and U.S. agencies' activities for prioritizing and coordinating U.S. interagency physical protection visits and measuring performance do not meet our best practices for agency performance or DOE's standards for internal control. We have reported that defining the mission and desired outcomes, measuring performance, and using performance information to identify performance gaps are critical if agencies are to be accountable for achieving intended results.[27] In addition, DOE's own standards for internal control call for "processes for planning, organizing, directing, and controlling operations designed to reasonably assure that programs achieve intended results... and decisions are based on reliable data."[28] However, DOE, NRC, and State have neither established a plan nor measured performance to determine whether they are meeting internal goals and whether U.S. agencies' activities are systematic.

DOE and U.S. Agencies Do Not Systematically Visit Countries with Category I U.S. Nuclear Material or Revisit Foreign Facilities Not Meeting Security Guidelines

U.S. agencies have not systematically evaluated the security of foreign facilities holding U.S. nuclear material in two key ways. First, U.S.

interagency physical protection teams have not systematically visited countries holding Category I quantities of U.S. nuclear material. Second, interagency teams have not revisited sites that did not meet IAEA security guidelines in a timely manner.

U.S. Physical Protection Teams Have Not Systematically Visited Countries Holding Category I Quantities of Nuclear Material

U.S. interagency physical protection teams have not systematically visited countries believed to be holding Category I quantities of U.S. special nuclear material at least once every 5 years—a key programmatic goal. In a December 2008 document, DOE officials noted that, in 1987, NRC recommended that countries possessing Category I nuclear material be revisited at least once every 5 years. This recommendation was adopted as a goal for determining the frequency of follow-on visits. In addition, DOE, NRC, and State officials told us that they aim to conduct physical protection visits at each country holding Category I quantities of U.S. nuclear material at least once every 5 years. We evaluated U.S. agencies' performance at meeting this goal by reviewing records of U.S. physical protection visits and other information. [29] We found that the United States had met this goal with respect to two countries by conducting physical protection visits at least once every 5 years since 1987 while they held Category I quantities of U.S. nuclear material. However, we estimated that 21 countries held Category I amounts of U.S. nuclear material during the period from 1987 through 2010 but were not visited once every 5 years while they held such quantities of U.S. nuclear material.

In addition, U.S. interagency physical protection teams have not visited all partner facilities believed to contain Category I quantities of U.S. special nuclear material to determine whether the security measures in place meet IAEA security guidelines. Specifically, we reviewed physical protection visit records and NMMSS data and identified 12 facilities that NMMSS records indicate received Category I quantities of U.S. HEU that interagency physical protection teams have never visited.

We identified four additional facilities that GTRI officials told us currently hold, and will continue to hold, Category I quantities of U.S. special nuclear material for which there is no acceptable disposition path in the United States. In addition, these facilities have not been visited by a U.S. interagency physical protection team, according to our review of available documentation.[30]

Moreover, U.S. interagency physical protection teams have not systematically visited partner storage facilities for U.S. nuclear material. The

AEA, as amended, requires that U.S. nuclear cooperation agreements contain a stipulation giving the United States approval rights over any storage facility containing U.S. unirradiated or "separated" plutonium or HEU. DOE and NRC officials told us there is no list of such storage facilities besides those listed in a U.S. nuclear cooperation agreement with a certain partner. They stated—and our review of available documents corroborated—that a number of the U.S. physical protection visits have included assessments of overseas storage sites for U.S. nuclear material, since such sites are often collocated with research reactors. However, our review also found two instances where partner storage areas containing U.S. HEU or separated plutonium did not meet IAEA guidelines or were identified as potentially vulnerable.

U.S. Teams Have Not Revisited Facilities That Did Not Meet IAEA Security Guidelines in a Timely Manner

DOE and U.S. agencies do not have a systematic process to revisit or monitor security improvements at facilities that do not meet IAEA security guidelines. Based on our analysis of available documentation, we found that, since 1994, U.S. interagency physical protection teams determined that partner country sites did not meet IAEA security guidelines on 21 visits. We then examined how long it took for a U.S. team to revisit the sites that did not meet IAEA security guidelines and found that, in 13 of 21 cases, U.S. interagency teams took 5 years or longer to revisit the facilities.[31]

According to DOE, NRC, and State officials, the interagency physical protection visits are not the only way to determine whether partner facilities are meeting IAEA security guidelines. For example, the United States is able to rely on information provided by other visits and U.S. embassy staff to monitor physical security practices. These visits include DOE-only trips and trips by DOE national laboratory staff and NRC physical protection experts who worked with the host country to improve physical security at the sites. NRC officials also stated that, in some cases, the partner's corrective actions at the site are verified by U.S. officials stationed in the country, and a repeat physical protection visit is not always required.

IAEA officials told us that U.S. technical experts often participate in voluntary IAEA physical security assessments at IAEA member states' facilities. Specifically, IAEA created the International Physical Protection Advisory Service (IPPAS) to assist IAEA member states in strengthening their national security regime. At the request of a member state, IAEA assembles a team of international experts who assess the member state's system of physical protection in accordance with IAEA security guidelines. As of December 2010, 49 IPPAS missions spanning about 30 countries had been completed.

DOE SEEKS TO INCREASE SECURITY OR REMOVE VULNERABLE U.S. NUCLEAR MATERIAL AT PARTNER FACILITIES BUT FACES CHALLENGES

DOE has taken steps to improve security at a number of facilities overseas that hold U.S. nuclear material. DOE's GTRI program removes nuclear material from vulnerable facilities overseas and has achieved a number of successes. However, DOE faces a number of constraints. Specifically, GTRI can only bring certain types of nuclear material back to the United States that have an approved disposition pathway and meet the program's eligibility criteria. In addition, obtaining access to the partner facilities to make physical security improvements may be difficult. There are a few countries that are special cases where the likelihood of returning the U.S. nuclear material to the United States is considered doubtful.

DOE's Office of Nonproliferation and International Security and GTRI officials told us that when a foreign facility with U.S.-obligated nuclear material does not meet IAEA security guidelines, the U.S. government's first response is to work with the partner country to encourage physical security improvements. In addition, the GTRI program was established in 2004 to identify, secure, and remove vulnerable nuclear material at civilian sites around the world and to provide physical protection upgrades at nuclear facilities that are (1) outside the former Soviet Union, (2) in non-weapon states, and (3) not in high-income countries. According to GTRI officials, the U.S. government's strategy for working with partner countries to improve physical security includes: (1) encouraging high-income countries to fund their own physical protection upgrades with recommendations by the U.S. government and (2) working with otherthan-high-income countries to provide technical expertise and funding to implement physical protection upgrades. If the material is excess to the country's needs and can be returned to the United States under an approved disposition pathway, GTRI will work with the country to repatriate the material.

According to GTRI officials, GTRI was originally authorized to remove to the United States, under its U.S. fuel return program, only U.S.-obligated fresh and spent HEU in Material Test Reactor fuel, and Training Research Isotope

General Atomics (TRIGA) fuel rod form. According to GTRI officials, GTRI has also obtained the authorization to return additional forms of U.S. fresh and spent HEU, as well as U.S. plutonium from foreign countries, so long as there is no alternative disposition path. The material must (1) pose a threat to national security, (2) be usable for an improvised nuclear device, (3) present a high-risk of terrorist theft, and (4) meet U.S. acceptance criteria.

To date, GTRI has removed more than 1,240 kilograms of U.S. HEU from Australia, Argentina, Austria, Belgium, Brazil, Canada, Chile, Colombia, Denmark, Germany, Greece, Japan, the Netherlands, Philippines, Portugal, Romania, Slovenia, South Korea, Spain, Sweden, Switzerland, Taiwan, Thailand, and Turkey.[32] It has also performed security upgrades at reactors containing U.S. nuclear material that were not meeting IAEA security guidelines in 10 partner countries. As we reported in September 2009,[33] GTRI has improved the security of research reactors, and GTRI officials told us in April 2011 that they plan to continue to engage other countries to upgrade security.

In a separate report published in December 2010, we noted that GTRI has assisted in the conversion from the use of HEU to LEU or verified the shutdown of 72 HEU research reactors around the world,[34] 52 of which previously used U.S. HEU. GTRI prioritizes its schedule for upgrading the security of research reactors and removing nuclear material based on the amount and type of nuclear material at the reactor and other threat factors, such as the vulnerability of facilities, country-level threat, and proximity to strategic assets. Our review identified several situations where GTRI or its predecessor program removed vulnerable U.S. nuclear material.

Notwithstanding these successes, the GTRI program has some limitations. GTRI cannot remove all potentially vulnerable nuclear material worldwide because the program's scope is limited to only certain types of material that meet the eligibility criteria. GTRI officials told us that, of the approximately 17,500 kilograms of HEU it estimates was exported from the United States, the majority—12,400 kilograms—is currently not eligible for return to the United States. According to GTRI officials, over 10,000 kilograms is contained in fuels from "special purpose" reactors that are not included in GTRI's nuclear material return program because they were not traditional aluminum-based fuels, TRIGA fuels, or target material. As a result, this material does not have an acceptable disposition pathway in the United States, according to GTRI officials. GTRI officials stated that these reactors are in Germany, France, and Japan, and that the material has been deemed to be adequately protected. GTRI reported that the other approximately 2,000 kilograms of transferred U.S. nuclear material is located primarily in EURATOM member countries and is either currently in use or adequately protected.

In addition, the potential vulnerability of nuclear material at certain high-income facilities was raised to us by officials at the National Security Council (NSC)—the President's principal forum for considering national security and foreign policy matters—and included in a prior report. Specifically, we reported that, there may be security vulnerabilities in certain high-income countries, including three specific high-income countries named by the NSC officials. For sites in these countries, GTRI officials told us the U.S. government's strategy is to work bilaterally with the countries and to provide recommendations to improve physical protection, and follow up as needed.

Our analysis of available agency physical protection visit documents also raises concerns regarding the physical security conditions in these countries, including facilities that did not meet IAEA security guidelines and interagency physical protection teams' lack of access issues.

DOE also works with countries to remove material if it is in excess of the country's needs and meets DOE acceptance criteria. The ability of DOE to return U.S. nuclear material depends, however, on the willingness of the foreign country to cooperate. As we reported in September 2009, [35] because GTRI's program for physical security upgrades and nuclear material returns is voluntary, DOE faces some challenges in obtaining consistent and timely cooperation from other countries to address security weaknesses. Our report further noted that DOE has experienced situations where a foreign government has refused its assistance to make security upgrades. For example, we reported that one country had refused offers of DOE physical security upgrades at a research reactor for 9 years. However, this situation was subsequently resolved when all HEU was removed from this country, according to GTRI officials. In addition, we reported that DOE had experienced two other situations where the partner country would not accept security assistance until agreements with the United States were reached on other issues related to nuclear energy and security.

There are several countries that have U.S. nuclear material that are particularly problematic and represent special cases. Specifically, U.S. nuclear material has remained at sites in three countries where physical protection measures are unknown or have not been visited by an interagency physical protection team in decades. GTRI removed a large quantity of U.S. spent HEU recently from one of these countries.

According to NRC and State officials, U.S. transfers to these three countries were made prior to 1978, when the physical protection requirements were added to the AEA. Therefore, these countries have not made the same commitments regarding physical security of U.S.- transferred material. Finally, we identified another country that poses special challenges. All U.S-obligated HEU has been removed from this country, which was one of the GTRI program's highest priorities. Previous U.S. interagency physical protection visits found a site in this country did not meet IAEA security guidelines.

CONCLUSION

The world today is dramatically different than when most U.S. nuclear cooperation agreements were negotiated. Many new threats have emerged, and nuclear proliferation risks have increased significantly. We recognize that the United States and its partners share a strong common interest in deterring and preventing the misuse of U.S. nuclear material— or any nuclear material—and that flexibility in the agreements is necessary to forge strong and cooperative working relationships with our partners. The fundamental question, in our view, is whether nuclear cooperation agreements and their underlying legislative underpinnings need to be reassessed given the weaknesses in inventory management and physical security that we identified.

Specifically, we found these agreements may not be sufficiently robust in two areas—inventories and physical security. Without an accurate inventory of U.S. nuclear materials—in particular, weapon-usable HEU and separated plutonium—the United States does not have sufficient assurances regarding the location of materials. As a result, the United States may not be able to monitor whether the partner country is appropriately notifying the United States and whether the United States is appropriately and fully exercising its rights of approval regarding the transfer, retransfer, enrichment and reprocessing and, in some cases, storage of nuclear materials subject to the agreement terms. NRC and multiple offices within DOE could not provide us with an authoritative list of the amount, location, and disposition of U.S. HEU or separated plutonium overseas. We are particularly concerned that NRC and DOE could not account, in response to a 1992 mandate by Congress, on the location and disposition of U.S. nuclear material overseas—and that they have not developed such an inventory in the almost two decades since that mandate.

We recognize that physical security is a national responsibility. We also recognize that neither the AEA, as amended, nor the U.S. nuclear cooperation agreements in force require that State negotiate new or renewed nuclear cooperation agreement terms that include specific access rights for the United States to verify whether a partner is maintaining adequate physical security of U.S. nuclear material. Without such rights, it may be difficult for the United States to have access to critical facilities overseas—especially those believed to be holding weapon-usable materials—to better ensure that U.S. material is in fact adequately protected while the material remains in the partner's custody. We note the agreements are reciprocal, with both parties generally agreeing to all conditions specified in them. We acknowledge that any change to the nuclear cooperation framework or authorizing legislation will be very sensitive. Careful consideration should be given to the impact of any reciprocity clauses on U.S. national security when negotiating or reviewing these agreements. However, it may be possible to do so in a way that includes greater access to critical facilities where weapon-usable U.S. nuclear material is stored, without infringing on the sovereign rights of our partners or hampering the ability of the U.S. nuclear industry to remain competitive.

In the course of our work, we identified several weaknesses in DOE, NRC, and State's efforts to develop and manage activities that ensure that U.S. nuclear cooperation agreements are properly implemented. Specifically, the lack of a baseline inventory of U.S. nuclear materials—in particular, weapon-usable materials—and annual inventory reconciliations with all partners limits the ability of the U.S. government to identify where the material is located. Currently, annual reconciliations with five partners are undertaken. However, the information, with the exception of one country, is aggregated and not provided on a facility-by-facility basis. Without such information on facilities, it may be difficult to track U.S. material for accounting and control purposes. No annual reconciliations currently exist for the United States' other partners that it has transferred material to or trades with. The NMMSS database could be the official central repository of data regarding U.S. inventories of nuclear material overseas if DOE and NRC are able to collect better data.

We are concerned that DOE has not worked with NRC and State to develop a systematic process for monitoring and evaluating the physical security of U.S. nuclear material overseas, including which foreign facilities to visit for future physical protection visits. In particular, U.S. interagency physical protection teams have neither met a key programmatic goal for visiting countries containing Category I quantities of U.S. special nuclear material every 5 years, nor have they visited all partner facilities believed to be holding Category I quantities of U.S. nuclear material, nor revisited facilities that were found to not meet IAEA security guidelines in a timely manner.

Moreover, relying on reported thefts of U.S. nuclear material as a gauge of security is not the best measure of program effectiveness when accounting processes for inventory of U.S. material at foreign facilities are limited. Improving the U.S. government's management of nuclear cooperation agreements could contribute to the administration achieving its goal of securing all vulnerable nuclear material worldwide in 4 years.

MATTERS FOR CONGRESSIONAL CONSIDERATION

- Congress may wish to consider directing DOE and NRC to complete a full accounting of U.S. weapon-usable nuclear materials—in particular, HEU and separated plutonium—with its nuclear cooperation agreement partners and other countries that may possess such U.S. nuclear material.
- In addition, Congress may wish to consider amending the AEA if State, working with other U.S. agencies, does not include enhanced measures regarding physical protection access rights in future agreements and renewed agreements, so that U.S. interagency physical protection teams may obtain access when necessary to verify that U.S. nuclear materials have adequate physical protection. The amendment could provide that the U.S. government may not enter into nuclear cooperation agreements unless such agreements contain provisions allowing the United States to verify that adequate physical security is exercised over nuclear material subject to the terms of these agreements.

RECOMMENDATIONS FOR EXECUTIVE ACTION

We are making seven recommendations to enable agencies to better account for, and ensure the physical protection of, U.S. nuclear material overseas.

To help federal agencies better understand where U.S. nuclear material is currently located overseas, we recommend that the Secretary of State, working with the Secretary of Energy and the Chairman of the Nuclear Regulatory Commission, take the following four actions to strengthen controls over U.S. nuclear material subject to these agreements:

- determine, for those partners with which the United States has transferred material but does not have annual inventory reconciliation, a baseline inventory of weapon-usable U.S. nuclear material, and establish a process for conducting annual reconciliations of inventories of nuclear material on a facility-by-facility basis;
- establish for those partners with which the United States has an annual inventory reconciliation, reporting on a facility-by-facility basis for weapon-usable material where possible;
- facilitate visits to sites that U.S. physical protection teams have not visited that are believed to be holding U.S. Category I nuclear material; and
- seek to include measures that provide for physical protection access rights in new or renewed nuclear cooperation agreements so that U.S. interagency physical protection teams may in the future obtain access when necessary to verify that U.S. nuclear materials are adequately protected. Careful consideration should be given to the impact of any reciprocity clauses on U.S. national security when negotiating or reviewing these agreements.

In addition, we recommend that the Secretary of Energy, working with the Secretary of State, and the Chairman of the Nuclear Regulatory Commission take the following three actions:

- develop an official central repository to maintain data regarding U.S. inventories of nuclear material overseas. This repository could be the NMMSS database, or if the U.S. agencies so determine, some other official database;
- develop formal goals for and a systematic process to determine which foreign facilities to visit for future interagency physical protection visits. The goals and process should be formalized and agreed to by all relevant agencies; and
- periodically review performance in meeting key programmatic goals for the physical protection program, including determining which countries containing Category I U.S. nuclear material have been visited within the last 5 years, as well as determining whether partner facilities previously found to not meet IAEA security guidelines were revisited in a timely manner.

AGENCY COMMENTS AND OUR EVALUATION

We provided a draft of this report to the Secretaries of Energy and State, and the Chairman of the NRC for their review and comment. All three agencies generally disagreed with our conclusions and recommendations. DOE, NRC, and State disagreed with GAO in three general areas of the report. Specifically, all the agencies (1) disagree with our recommendations to establish annual inventory reconciliations with all trading partners and establish a system to comprehensively track and account for U.S. nuclear material overseas, because the agencies believe this is impractical and unwarranted; (2) maintain that IAEA safeguards are sufficient or an important tool to account for U.S. nuclear material overseas; and (3) assert that any requirement in future nuclear cooperation agreements calling for enhanced physical protection access rights is unnecessary and could hamper sensitive relationships.

With regard to the three general areas of disagreement, our response is as follows:

- DOE, NRC, and State assert that it is not necessary to implement GAO's recommendation that agencies undertake an annual inventory reconciliation and report on a facility-by-facility basis for weapon-usable material where possible for all countries that hold U.S.-obligated nuclear material. We stand by this recommendation for numerous reasons. First, as stated in the report, we found—and none of the agencies refuted—that the U.S. government does not have an inventory of U.S. nuclear material overseas and, in particular, is not able to identify where weapon-usable materials such as HEU and separated plutonium that can be used for a nuclear weapon may reside. In fact, NRC commented that "inventory knowledge is very important for high-consequence materials, e.g., high enriched uranium and separated plutonium." Because DOE, NRC, and State do not have comprehensive knowledge of where U.S.-obligated material is located at foreign facilities, it is unknown whether the United States is appropriately and fully exercising its rights of approval regarding the transfer, retransfer, enrichment, and reprocessing and, in some cases, storage of nuclear materials subject to the agreements' terms. In addition, the lack of inventory information hampers U.S. agencies in identifying priorities for interagency physical protection visits. We are particularly concerned that NRC and DOE, in response to a 1992 mandate by Congress, could only account for the location and disposition of about 1,160 kilograms out of an estimated 17,500 kilograms of U.S.-exported HEU. Furthermore, the agencies have not developed such an inventory or performed an additional comprehensive review in the almost two decades since that mandate. We believe it is important that DOE, NRC, and State pursue all means possible to better identify where U.S.-obligated material is located overseas—and for weapon-usable HEU and separated plutonium, seek to do so on a facility-by-facility basis. Annual inventory reconciliations with all partners provide one way to do that. The United States has demonstrated it has the ability to conduct such exchanges, which none of the agencies disputed. Our report notes that the United States conducts annual inventory reconciliations with five partners, including one where facility-level information is annually exchanged. We believe the recent signing of nuclear cooperation agreements with India and Russia, as well as the situation where current partners whose agreements are set to expire in coming years must be renegotiated—including Peru and South Korea—provide a convenient and timely opportunity for DOE, NRC, and State to pursue such enhanced material accountancy measures.

- DOE, NRC, and State commented that IAEA's comprehensive safeguards program is another tool to maintain the knowledge of locations of nuclear material in a country, including U.S.-obligated material, and that IAEA inspection, surveillance, and reporting processes are effective tools for material tracking and accounting. We agree that IAEA safeguards are an important nuclear nonproliferation mechanism. However, our report found IAEA's safeguards have a limited ability to identify, track, and account for U.S.-obligated material. Specifically, as our report notes, and as confirmed to us by senior IAEA officials, IAEA does not track the obligation of the nuclear material under safeguards and, therefore, IAEA may not have the ability to identify whether and what volume of nuclear material at partner country facilities is U.S.-obligated and subject to the terms of U.S. nuclear cooperation agreements. In addition, our report notes that IAEA considers member country nuclear material inventory information confidential and does not share it with its member countries, including the United States. Therefore, IAEA has a limited ability to account for nuclear material subject to the terms of U.S. nuclear cooperation agreements. Importantly, safeguards are not a substitute for physical security and serve a different function. As our report notes, safeguards are primarily a way to detect diversion of nuclear material from peaceful to military purposes but do not ensure that facilities are physically secure to prevent theft or sabotage of such material.

- DOE, NRC, and State disagreed with our recommendation that State, working with DOE and NRC, should seek to negotiate terms that include enhanced measures regarding physical protection access rights in future and renewed agreements. They also raised concerns with our Matter for Congressional Consideration to amend the AEA should State not implement our recommendation. We do not agree with agencies' comments that our recommendation that agencies "seek to include" such measures is impractical. As we note in our report, an enhanced measure for access rights is in place in the recently negotiated U.S.-India arrangements and procedures document. Further, while partner countries pledge at the outset of an agreement that they will physically protect U.S.- obligated material, the results of our work show that they have not always adequately done so. Specifically, our report noted that, of the 55 interagency physical protection visits made from 1994 through 2010, interagency teams found that countries met IAEA security guidelines on only 27 visits; did not meet IAEA security guidelines on 21 visits, and the results of 7 visits are unknown because the U.S. team was unable to assess the sites or agency documentation of the physical protection visits was missing. In addition, we identified 12 facilities that are believed to have or previously had Category I U.S. nuclear material that have not been visited by an interagency physical protection team. We agree with the agencies' comments that the licensing process for U.S. nuclear material offers some assurances that physical security will be maintained and that an exchange of diplomatic notes at the time of a transfer is designed to ensure the partners maintain the material according to the terms of the agreements. However, these measures are implemented at the time of licensing or material nuclear material over the longer-term, often 30-year duration of these agreements is by no means guaranteed. Ensuring that the United States has the tools it needs to visit facilities in the future—even after an initial transfer of material is made per a conditional export license—is important to supporting U.S. nuclear nonproliferation objectives. We continue to believe that our recommendation and Matter for Congressional

Consideration are consistent with the report's findings and would enhance the security of U.S.-obligated nuclear material in other countries.

In addition, DOE and NRC commented that (1) our report contained errors in fact and judgment, (2) our report's recommendations could result in foreign partners requiring reciprocal access rights to U.S. facilities that contain nuclear material that they transferred to the United States, which could have national security implications, and (3) our recommendation that agencies establish a process for conducting annual reconciliations of inventories of nuclear material and develop a repository to maintain data regarding U.S. inventories of nuclear material overseas would be costly to implement. Our response to these comments is as follows:

- None of the agencies' comments caused us to change any factual statement we made in the report. DOE provided a limited number of technical comments, which we incorporated as appropriate. Importantly, some of the facts that agencies did not dispute included: (1) our analysis that found U.S. agencies made only a single attempt to comprehensively account for transferred U.S. HEU almost 20 years ago and, at that time, were only able to verify the amount and location of less than one-tenth of transferred U.S. HEU; and (2) partner countries did not meet IAEA physical security guidelines for protecting U.S. nuclear material in about half of the cases we reviewed from 1994 through 2010. In our view, these security weaknesses place U.S.-obligated nuclear material at risk and raise potential proliferation concerns. These agreements for nuclear cooperation are long-term in scope and are often in force for 30 years or more. As we noted in our report, the world today is dramatically different than the time when most of the agreements were negotiated. New threats have emerged, and nuclear proliferation risks have increased significantly. NRC commented that countries may not want to change the "status quo" as it pertains to nuclear cooperation agreement terms, including those regarding the physical protection of U.S.-obligated nuclear material. In our view, the status quo, or business-as-usual approach should not apply to matters related to the security of U.S.-obligated nuclear material located at partner facilities throughout the world. Moreover, implementing a more robust security regime is consistent with and complements the administration's goal of securing all vulnerable nuclear material worldwide within a 4-year period.

- DOE and NRC's comment that the United States may be asked to demonstrate reciprocity by nuclear cooperation agreement partners to verify that adequate physical protection is being provided to their nuclear material while in U.S. custody has merit and needs to be taken into consideration when developing or reviewing nuclear cooperation agreements. As a result, we added language to the conclusions and recommendation sections to additionally state that "careful consideration should be given to the impact of any reciprocity clauses on U.S. national security when negotiating or reviewing these agreements."
- In addition, DOE and NRC commented that we are suggesting a costly new effort in recommending that agencies account for and track U.S.-obligated nuclear material overseas. However, we noted in our report that NMMSS officials told us that NMMSS is currently capable of maintaining information regarding inventories of U.S. nuclear material overseas. Moreover, DOE and NRC did not conduct an analysis to support their assertion that such a system would be costly. Although we did not perform a cost-benefit analysis, based on our conversations with NMMSS staff and the lack of a DOE cost-benefit analysis, to the contrary, there is no evidence to suggest that adding additional information to the NMMSS database would necessarily entail significant incremental costs or administrative overhead. We are sensitive to suggesting or recommending new requirements on federal agencies that may impose additional costs. However, it is important to note that the U.S. government has already spent billions of dollars to secure nuclear materials overseas, as well as radiation detection equipment to detect possible smuggled nuclear material at our borders and the border crossings of other countries. The administration intends to spend hundreds of millions more to support the president's 4-year goal to secure all vulnerable nuclear material worldwide. If necessary, an expenditure of some resources to account for U.S. nuclear material overseas is worthy of consideration. We stand by our recommendations that State work with nuclear cooperation agreement partners that the United States has transferred material to, to develop a baseline inventory of U.S. nuclear material overseas, and that DOE work with other federal

agencies to develop a central repository to maintain data regarding U.S. inventories of nuclear material overseas.

DOE disagreed with our findings that the U.S. interagency physical protection visit program (1) lacked formal goals, and that (2) U.S. agencies have not established a formal process for coordinating and prioritizing interagency physical protection visits, in addition to the three areas of general disagreement. During the course of our work, we found no evidence of an interagency agreed-upon list of program goals. In its comments, DOE stated that the formal goal of the program is to determine whether U.S.-obligated nuclear material at the partner country facility is being protected according to the intent of IAEA security guidelines. This is the first time the goal has been articulated to us as such. Moreover, we disagree with DOE's second assertion that it has established a formal process for coordinating and prioritizing visits. Our report notes that we found DOE has not (1) worked with NRC and State to establish a plan and prioritize U.S. physical protection visits or (2) measured performance in a systematic way. In particular, our report notes that, in October 2009, a DOE Office of Nonproliferation and International Security official reported to us that it had formulated a list of 10 countries that contained U.S. nuclear material and were priorities for physical protection teams to visit. However, a senior-level DOE nonproliferation official told us that DOE had not discussed this list with State or NRC, or other agency officials, and it could not be considered an interagency agreed-upon list. In addition, NRC Office of International Program officials told us they thought interagency coordination could be improved, and a State Bureau of International Security and Nonproliferation official told us that agency coordination has improved in the past 6 months. Moreover, as we further state in the report, in February 2011, DOE officials told us that the department is conducting a study of its methodology for prioritizing physical protection visits.

In addition, in July 2011, in conjunction with the classification review for this report, DOE officials stated that while DOE, NRC, and State work together on coordinating U.S. government positions regarding priorities and procedures for the interagency physical protection program, no updated document exists that formalizes the process for planning, coordinating, and prioritizing U.S. interagency physical protection visits. We note that the documents that DOE refers to are internal DOE documents presented to GAO in 2008 and 2009 in response to questions regarding nuclear cooperation agreements. These documents are not an interagency agreed upon document, but reflects DOE's views on determining which countries and facilities interagency physical protection teams should visit. Further, DOE officials in July 2011 stated that DOE, NRC, and State do not have an agreed-upon way to measure performance in a systematic way, and that while the goals for the monitoring and evaluation activities have not yet been formalized through necessary updated documents, a prioritized list of countries to visit does exist. These officials noted that the U.S. government is working to update its planning documents and examining its methodology for prioritizing physical protection visits. Any changes will be included in these updated documents. Therefore, we continue to believe that DOE should work with the other agencies to develop formal goals for and a systematic process for determining which foreign facilities to visit for future physical protection visits, and that the process should be formalized and agreed to by all agencies.

NRC commented that in order to demonstrate that U.S. nuclear material located abroad is potentially insecure, GAO made an assessment based on U.S. agencies not conducting activities which are, according to NRC, neither authorized nor required by U.S. law or by agreements negotiated under Section 123 of the AEA. In fact, we acknowledge that U.S. agencies are not required to conduct certain activities or collect certain information. Moreover, we do not suggest that agencies undertake activities that are not authorized by law. We recommend that the agencies either expand upon and refine outreach they are already conducting, contingent on the willingness of our cooperation agreement partners, or negotiate new terms in nuclear cooperation agreements as necessary. If the agencies find that they are unable to negotiate new terms we recommend that Congress consider amending the AEA to require such terms.

State commented that determining annual inventories and reconciliations of nuclear material, as well as establishing enhanced facility-by-facility reporting for those partners with which the United States already has an annual inventory reconciliation is a DOE function, not a State function. We agree that DOE plays a vital role in carrying out these activities— once such bilaterally agreed upon measures are in place. However, we believe it is appropriate to recommend that the Department of State—as the agency with the lead role in any negotiation regarding the terms and conditions of U.S. nuclear cooperation agreements—work with DOE and NRC to secure these measures with all U.S. partners. State also commented that there is a cost to the U.S. nuclear industry in terms of lost competitiveness should the requirements in U.S. nuclear cooperation agreements be strengthened to include better access to critical facilities for U.S. interagency physical protection teams. State provided no further information to support this point. Our report acknowledges that any change to the nuclear cooperation framework or authorizing legislation will be very sensitive and that flexibility in the agreements is necessary.

We also stated that it may be possible to change the framework of agreements in a way that does not hamper the ability of the U.S. nuclear industry to remain competitive. While we would not want to alter these agreements in such a way that our nuclear industry is put at a competitive disadvantage, in our view, the security of U.S. nuclear material overseas should never be compromised to achieve a commercial goal.

Finally, State asserted that interagency physical protection teams have been granted access to every site they have requested under the consultation terms of U.S. nuclear cooperation agreements. As a result, State believes the provisions of the current agreements are adequate. As we note in our report, access to partner facilities is not explicitly spelled out in the agreements and, in our view, this is a limitation for the U.S. agencies in obtaining timely and systematic access to partner nuclear facilities. While State may be technically correct that access has been granted, our report clearly shows that many sites believed to contain Category I quantities of U.S. nuclear material have been visited only after lengthy periods of time, or have not been visited at all. We continue to believe that enhanced physical protection access measures could help interagency teams ensure that they are able to visit sites containing U.S. nuclear material in a timely, systematic, and comprehensive fashion.

Gene Aloise
Director,
Natural Resources and Environment

APPENDIX I: OBJECTIVES, SCOPE, AND METHODOLOGY

We addressed the following questions during our review: (1) assess U.S. agency efforts to account for U.S. nuclear material overseas, (2) assess the Department of Energy's (DOE) and other U.S. agencies' efforts to monitor and evaluate the physical security conditions of U.S. nuclear material subject to the terms of nuclear cooperation agreements, and (3) describe DOE's activities to secure or remove potentially vulnerable U.S. nuclear material at partner facilities.

To assess U.S. agency efforts to account for U.S. nuclear material overseas, we reviewed relevant statutes, including the Atomic Energy Act of 1954 (AEA), as amended, as well as the texts of all current nuclear cooperation agreements. We obtained data from the Nuclear Materials Management and Safeguards System (NMMSS), a database jointly run by DOE and the Nuclear Regulatory Commission (NRC), which, among other things, maintains data on U.S. peaceful use exports and retransfers of enriched uranium and plutonium that have occurred since 1950, and reviewed DOE and GAO reviews of the NMMSS database. To assess the reliability of data in the NMMSS database, we interviewed officials from DOE and NRC and a former DOE contractor to identify any limitations in NMMSS's data on the location and status of U.S. material overseas and found these data to be sufficiently reliable for the purposes of accounting for U.S. exports of nuclear material. We compared NMMSS data with other official and unofficial DOE sources of information regarding U.S. nuclear material transfers, including DOE data on nuclear material returns, to determine the reliability of DOE's inventory data for U.S. nuclear material transferred overseas. We reviewed DOE, NRC, and other U.S. agency records and interviewed officials at those agencies to determine the extent to which DOE, NRC, and State are able to identify where U.S. nuclear material was exported, retransferred, and is currently held. We selected a non-probability sample of partners based on, among other considerations, quantities of U.S. special nuclear material transferred to them. Results of interviews of non-probability samples are not generalizeable to all partners but provide an understanding of those partners' views of the U.S. government's efforts to account for its nuclear material inventories overseas subject to nuclear cooperation agreement terms. We conducted site visits in four countries holding U.S.-obligated material and interviewed governmental officials and nuclear facility operators in these countries to discuss material accounting procedures. Further, we interviewed officials from five partners regarding their observations about working with the U.S. government to account for material subject to the terms of nuclear cooperation agreements. We analyzed the texts of administrative arrangements with key countries to determine the extent to which DOE conducts inventory reconciliations of inventory transferred between the United States and a partner country.

To assess DOE's and other U.S. agencies' efforts to monitor and evaluate the physical security conditions of U.S. nuclear material overseas subject to nuclear cooperation agreement terms and describe DOE's activities to secure or remove potentially vulnerable U.S. nuclear material at partner facilities, we reviewed all U.S. nuclear cooperation agreements in force, as well as other U.S. statutes, and IAEA's security guidelines, "The Physical Protection of Nuclear Material and Nuclear Facilities," INFCIRC/225/Rev.4,1 and other relevant international conventions to determine the extent to which such laws and international conventions provide for DOE and U.S. agencies to monitor and evaluate the physical security of transferred U.S. nuclear material subject to U.S. nuclear cooperation agreement terms. We interviewed officials from DOE, NRC, and the Department of State (State) to gain insights into how effective their efforts are, and how their efforts might be improved. We selected a nonprobability sample of partners based on, among other considerations, quantities of U.S. special nuclear material transferred to them and interviewed officials to determine how DOE and other U.S. agencies work with partner countries to exchange views on physical security and the process by which U.S. nuclear material is returned to the United States. Results of interviews of non-probability samples are not generalizeable to all partners but provide an understanding of those partners' views of the U.S. government's efforts to monitor and evaluate the physical security conditions of U.S. nuclear material overseas subject to nuclear cooperation agreement terms. We also obtained and analyzed the records of all available U.S. physical protection visits to partner facilities from 1974 through 2010. We reviewed agency documents and interviewed officials from DOE, NRC, and State regarding the policies and procedures for determining which partners to visit, how they conducted physical protection visits at partner facilities, and mechanisms for following up on the results of these visits. In particular, we compared the sites visited with NMMSS records of U.S. material exported and retransferred, and other information to evaluate the extent to which U.S. physical protection visits were made to all sites overseas containing U.S. special nuclear material. We obtained written responses from Global Threat Reduction Initiative (GTRI), and reviewed other information regarding their program activities. To better understand IAEA's role in maintaining safeguards and evaluating physical security measures, we interviewed IAEA officials and reviewed relevant documents.

We conducted this performance audit from September 2010 to June 2011 in accordance with generally accepted government auditing standards. Those standards require that we plan and perform the audit to obtain sufficient, appropriate evidence to provide a reasonable basis for our findings and conclusions based on our audit objectives. We believe that the evidence obtained provides a reasonable basis for our findings and conclusions based on our audit objectives.

APPENDIX II: CURRENT AND PREVIOUS U.S. NUCLEAR COOPERATION AGREEMENT PARTNERS

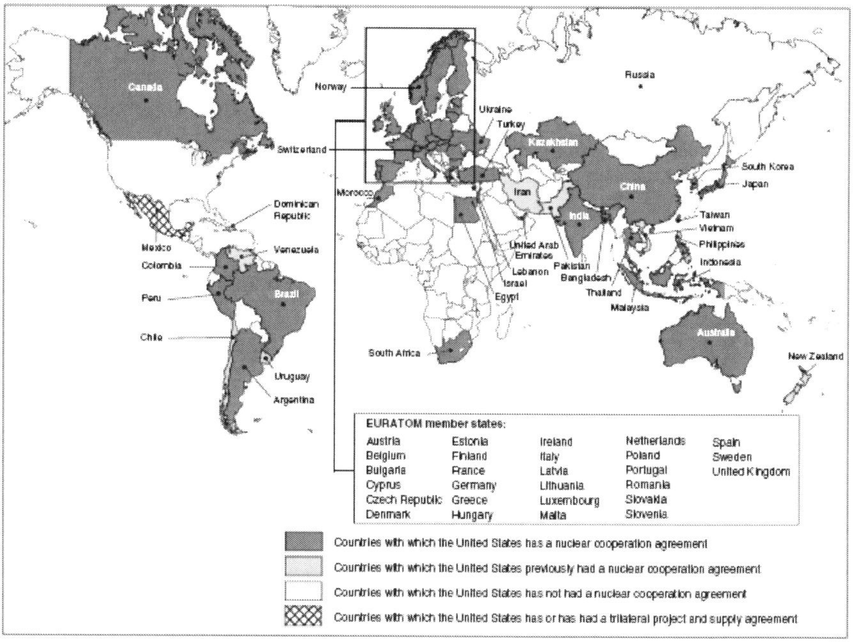

Sources: GAO analysis of Department of State data; Map Resources (map).

Figure 1. Cooperating Partners with Which the United States Currently Has or Previously Had a Nuclear Cooperation Agreement.

As indicated in figure 1, the United States has nuclear cooperation agreements in force with Argentina, Australia, Bangladesh, Brazil, Canada,

China, Colombia, EURATOM, Egypt, India, Indonesia, IAEA, Japan, Kazakhstan, Morocco, Norway, Peru, Russia, South Africa, South Korea, Switzerland, Taiwan, Thailand, Turkey, Ukraine, and United Arab Emirates.[1] In addition, the United States previously had nuclear cooperation agreements with Chile, Dominican Republic, Iran, Israel, Lebanon, New Zealand, Pakistan, Philippines, Uruguay, Venezuela, and Vietnam.[2]

APPENDIX III: INTERNATIONAL GUIDELINES FOR THE CATEGORIZATION OF NUCLEAR MATERIAL

IAEA's INFCIRC/225/Rev.4 security guideline document establishes the standard by which the United States and other countries generally classify the categories of physical protection that should be afforded nuclear material, based on the type, volume, and disposition of the nuclear material. Table 1 lists the material category according to IAEA's security guidelines, INFCIRC/225/Rev4. Specifically:

Table 1. Categorization of Nuclear Material According to IAEA Security Guidelines

Material	*Form*	*Category I*	*Category II*	*Category III*[c]
Plutonium[a]	Unirradiated[b]	2 kilograms or more	Less than 2 kilograms but more than 500 g	500 g or less but more than 15 g
Uranium-235	Unirradiatedb - uranium enriched to 20% 235U or more -uranium enriched to 10%235U but less than 20%235U - uranium enriched above natural, but less than 10%235U	5 kilograms or more	Less than 5 kilograms but more than 1 kilogram 10 kilograms or more	1 kilogram or less but more than 15g Less than 10 kilograms but more than 1 kilogram 10 kilograms or more
Uranium-233	Unirradiated[b]	2 kilograms or more	Less than 2 kilograms but more than 500 g	500 g or less but more than 15 g

Table 1. (Continued)

Material	*Form*	*Category I*	*Category II*	*Category III*[c]
4. Irradiated fuel (The categorization of irradiated fuel in the table is based on international transport considerations. The State may assign a different category for domestic use, storage, and transport taking all relevant factors into account.)			Depleted or natural uranium, thorium, or low-enriched fuel (less than 10% fissile content)[d,e]	

Source: IAEA INFCIRC225r4.

[a] All plutonium except that with isotopic concentration exceeding 80 % in plutonium-238.

[b] Material not irradiated in a reactor or material irradiated in a reactor but with a radiation level equal to or less than 1 Gy/hr (100 rad/hr) at 1 meter unshielded.

[c] Quantities not falling in Category III and natural uranium, depleted uranium and thorium should be protected at least in accordance with prudent management practice.

[d] Although this level of protection is recommended, it would be open to States, upon evaluation of the specific circumstances, to assign a different category of physical protection.

[e] Other fuel which by virtue of its original fissile material content is classified as Category I or II before irradiation may be reduced one category level while the radiation level from the fuel exceeds 1 Gy/hr (100 rad/hr) at 1 meter unshielded.

End Notes

[1] For the purposes of this report, we refer to all countries, EURATOM, IAEA, and Taiwan as partners. EURATOM is composed of the 27 countries of the European Union. IAEA, an independent international organization based in Vienna, Austria, is affiliated with the United Nations and has the dual mission of promoting the peaceful uses of nuclear energy and verifying that nuclear materials intended for peaceful purposes are not diverted to military purposes. IAEA had 151 member states as of November 2010. Governmental relations between the United States and Taiwan were terminated on January 1, 1979. All agreements concluded with the authorities on Taiwan prior to January 1, 1979, are administered for the United States by the American Institute in Taiwan, a nonprofit corporation based in Washington, D.C. The United States has two nuclear cooperation agreements with Australia, including one for Separation of Uranium Isotopes by Laser Excitation (SILEX) technology, bringing the number of agreements to 27.

[2] Special nuclear material includes uranium enriched in the isotope of uranium-235, uranium-233, and plutonium.

[3] U.S. companies can obtain a license from the Nuclear Regulatory Commission (NRC) to export certain small quantities of nuclear material and minor reactor components to foreign

countries without a nuclear cooperation agreement in place. Dual-use items— items that can be used for both civilian and military applications—can be exported without a nuclear cooperation agreement.

[4] GAO, Nuclear Nonproliferation: Concerns With the U.S. International Nuclear Materials Tracking System, GAO/T-RCED/AIMD-96-91 (Washington, D.C.: Feb. 28, 1996).

[5] Weapon-usable nuclear materials are HEU—uranium enriched in the isotope uranium-235 to 20 percent or greater; uranium-233; and any plutonium containing less than 80 percent of the isotope plutonium-238. Such materials are also often referred to as fissile materials or strategic special nuclear materials. In addition, weapon-grade HEU is generally defined as HEU enriched in the isotope of uranium-235 at 90 percent or greater.

[6] GAO, Nuclear Nonproliferation: Comprehensive U.S. Planning and Better Foreign Cooperation Needed to Secure Vulnerable Nuclear Materials Worldwide, GAO-11-227 (Washington, D.C.: Dec. 15, 2010).

[7] NNSA was created by the National Defense Authorization Act for Fiscal Year 2000, Pub. L. No. 106-65 (1999). It is a separate semiautonomous agency within DOE, with responsibility for the nation's nuclear weapons, nonproliferation, and naval reactors programs.

[8] GAO, Nuclear Commerce: Governmentwide Strategy Could Help Increase Commercial Benefits from U.S. Nuclear Cooperation Agreements with Other Countries, GAO-11-36 (Washington, D.C.: Nov. 4, 2010).

[9] Results of interviews of non-probability samples are not generalizeable to all partners but provide an understanding of those partners' views of the U.S. government's efforts to account for its nuclear material inventories and monitor and evaluate the physical security conditions of U.S. nuclear material overseas subject to nuclear cooperation agreement terms.

[10] In January 2011, IAEA issued an updated revision of its security guideline document, IAEA, "Nuclear Security Recommendations on Physical Protection of Nuclear Material and Nuclear Facilities" (IAEA INFCIRC/225/Rev.5 (2011).

[11] The President may exempt proposed agreements for peaceful nuclear cooperation from any of these requirements if he determines that the requirement would be seriously prejudicial to the achievement of U.S. nonproliferation objectives or otherwise jeopardize the common defense and security.

[12] Section 104 of the Henry J. Hyde United States-India Peaceful Atomic Energy Cooperation Act of 2006 authorized the President to exempt India from the full-scope safeguards requirement if the President made a determination that India had taken certain specified actions and that the Nuclear Suppliers Group (NSG) had decided by consensus to permit the supply to India of nuclear items covered by the NSG Guidelines. On September 10, 2008, the President made the requisite determination, the agreement was approved by legislation enacted on October 8, 2008, and that agreement with India entered into force on December 6, 2008.

[13] A production facility is any equipment or device, or any important component of such equipment or device, capable of the production of special nuclear material in such quantity as to be of significance to the common defense and security or in such a manner as to affect the health and safety of the public. A utilization facility is any equipment or device, or any important component of such equipment or device, other than an atomic weapon, capable of using such material.

[14] The U.S.-China agreement states that parties shall exchange views on their national material accounting systems.

[15] Energy Policy Act of 1992, Pub. L. No. 102-486, § 903(b), 106 Stat. 2776, 2945-46.

[16] NRC, The United States Nuclear Regulatory Commission's Report to Congress on the Disposition of Highly Enriched Uranium Previously Exported from the United States, Washington, D.C. (January 1993).

[17] NMMSS has been used to account for U.S. imports and exports of nuclear material since 1967 and has been upgraded several times, most recently in 2009, though some of its information dates to 1950.

[18] The United States has two agreements with Australia. One agreement concerns broad-based nuclear cooperation; the second is limited to collaboration in SILEX technology.

[19] According to DOE, a sealed source may contain nuclear or radiological material, and is packaged to be environmentally safe and are generally used for calibration of radiation measuring and monitoring instruments in nuclear research and development.

[20] See GAO, Obstacles To U.S. Ability to Control And Track Weapons-Grade Uranium Supplied Abroad, GAO/ID-82-91 (Washington, D.C. Aug. 2, 1982).

[21] DOE Draft Order O 470.6 Nuclear Material Control and Accountability, issued Oct. 6, 2010.

[22] An arrangements and procedures document negotiated pursuant to the 2008 U.S.-India nuclear cooperation agreement contains a provision providing for consultation visits at two reprocessing facilities established to reprocess material including U.S.-obligated material. In addition, the 1988 U.S.-Japan nuclear cooperation agreement provides for what is known as "advance consent rights" by the United States to Japan to reprocess its U.S.- obligated spent nuclear fuel. Specifically, a provision in the nuclear cooperation agreement allows for either party to "have access to all places and data, and any equipment or facility...necessary to account for the nuclear material...and to make such independent measurements as may be deemed necessary by the safeguarded party to account for such nuclear material" if either party becomes aware that the IAEA is not applying safeguards as required by the agreement.

[23] We reported in August 1982 and December 1994 that the United States evaluates foreign countries physical protection systems under what was then known as the U.S. Bilateral Physical Protection Program. For a list of countries visited, number of facilities visited, and date of visits from 1974 through 1981, see GAO/ID-82-81. In 1994, we reported that the United States had conducted bilateral physical security consultations with approximately 46 nations, including site visits to review the physical protection at fixed sites and during transport. For a list of the countries visited by U.S. physical protection teams from 1974 through 1994, including country, number of visits, and date of last visit, see appendix II of that report, GAO, Nuclear Nonproliferation: U.S. International Nuclear Materials Tracking Capabilities Are Limited, GAO/RCED/AIMD-95-5 (Washington, D.C.: Dec. 27, 1994).

[24] NRC's regulations pertaining to the review of license applications for exports of nuclear equipment, material including exports of material subject to nuclear cooperation agreements, permit the determination of adequacy of foreign protection systems on a countrywide basis. See 10 C.F.R. §110.44 (2011).

[25] LEU is uranium that contains less than 20 percent of the isotope uranium-235.

[26] Our analysis shows three countries were visited between October 2009 and April 2010, one more in early 2011, and DOE and State officials reported they expect to visit one other in late 2011.

[27] GAO, Executive Guide: Effectively Implementing the Government Performance and Results Act, GAO/GGD-96-118 (Washington, D.C.: June 1996).

[28] DOE Order O 413.1B.

[29] Because DOE and NRC did not have a comprehensive inventory of U.S. material located overseas or any comprehensive analyses of the U.S. interagency physical protection visits'

results—as described elsewhere in this report—we obtained and analyzed NMMSS records of U.S. nuclear material exports and retransfers, GTRI records of fuel returns and, where available, information from U.S. physical protection visits records indicating the volumes and disposition of U.S. nuclear material overseas at the time of the U.S. interagency visit. We developed an estimate of which partner countries held Category I quantities of U.S. nuclear material during the review period, and for how long. We used the IAEA security document INFCIRC/225/Rev.4 to categorize material. We shared the results of this analysis with DOE and NRC officials and made adjustments based on agency officials' comments on where they believe nuclear material subject to U.S. nuclear cooperation agreement terms resides.

[30] GTRI reported to us that these sites are not a physical security priority as, compared with other facilities around the world, their physical protection is adequate.

[31] In three cases, GTRI, or its predecessor organization, returned the material determined to be vulnerable within 5 years from the date of assessment by the U.S. interagency physical protection team. In two cases, a U.S. physical protection team made a revisit in less than 5 years. In the most recent two cases, 5 years have not elapsed to make a determination on whether the team visited within 5 years or not. We chose 5 years to evaluate because U.S. agencies have a goal of visiting countries with Category I nuclear material at least once every 5 years.

[32] GTRI has removed all U.S.-obligated HEU from Brazil, Chile, Colombia, Denmark, Greece, Philippines, Portugal, Romania, Slovenia, South Korea, Spain, Sweden, Taiwan, Thailand, and Turkey.

[33] GAO, Nuclear Nonproliferation: National Security Administration Has Improved the Security of Reactors in its Global Research Reactor Program, but Action is Needed to Address Remaining Concerns, GAO-09-949 (Washington, D.C.: Sept. 17, 2009).

[34] See GAO-11-227.

[35] See GAO-09-949.

End Note for Appendix I

[1] In January 2011, IAEA issued an updated revision of its security guideline document, IAEA, "Nuclear Security Recommendations on Physical Protection of Nuclear Material and Nuclear Facilities" (IAEA INFCIRC/225/Rev.5 (2011).

End notes for Appendix II

[1] The United States has a set of trilateral project and supply agreements with Mexico and IAEA. We included these agreements because they were entered into pursuant to the United State's nuclear cooperation agreement with IAEA. The United States has two nuclear cooperation agreements with Australia, including one for Separation of Uranium Isotopes by Laser Excitation (SILEX) technology, bringing the number of agreements to 27.

[2] The United States also previously had trilateral project and supply agreements with Malaysia, Yugoslavia, and IAEA.

In: Security of United States' Nuclear … ISBN: 978-1-62417-038-6
Editors: R. E. Allen and N. L. Miller

Chapter 2

NUCLEAR COOPERATION WITH OTHER COUNTRIES: A PRIMER*

Paul K. Kerr and Mary Beth Nikitin

SUMMARY

In order for the United States to engage in civilian nuclear cooperation with other states, it must conclude a framework agreement that meets specific requirements under Section 123 of the Atomic Energy Act (AEA). The AEA also provides for exemptions to these requirements, export control licensing procedures, and criteria for terminating cooperation. Congressional review is required for Section 123 agreements; the AEA establishes special parliamentary procedures by which Congress may act on a proposed agreement.

WHAT IS A "SECTION 123" AGREEMENT?

Under existing law (Atomic Energy Act [AEA] of 1954, as amended; P.L. 95-242; 42 U.S.C. §2153 et seq.)[1] all significant U.S. nuclear cooperation with other countries requires a peaceful nuclear cooperation agreement.[2]

---

* This is an edited, reformatted and augmented version of a Congressional Research Service publication, CRS Report for Congress RS22937, from www.crs.gov, prepared for Members and Committees of Congress, dated June 19, 2012.

Significant nuclear cooperation includes the transfer of U.S.- origin special nuclear material[3] subject to licensing for commercial, medical, and industrial purposes.

Such agreements, which are "congressional-executive agreements" requiring congressional approval, do not guarantee that cooperation will take place or that nuclear material will be transferred, but rather set the terms of reference and authorize cooperation. The AEA includes requirements for an agreement's content, conditions for the President to exempt an agreement from those requirements, presidential determinations and other supporting information to be submitted to Congress, conditions affecting the implementation of an agreement once it takes effect, and procedures for Congress to consider and approve the agreement.

Section 123 of the AEA requires that any agreement for nuclear cooperation meet nine nonproliferation criteria and that the President submit any such agreement to the House Committee on Foreign Affairs and the Senate Committee on Foreign Relations.

The Department of State is required to provide the President an unclassified Nuclear Proliferation Assessment Statement (NPAS), which the President is to submit, along with the agreement, to those two committees. The State Department is also required to provide a classified annex to the NPAS, prepared in consultation with the Director of National Intelligence.

The NPAS is meant to explain how the agreement meets the AEA nonproliferation requirements. The President must also make a written determination "that the performance of the proposed agreement will promote and will not constitute an unreasonable risk to, the common defense and security."

Requirements Under the Atomic Energy Act

Section 123 of the AEA specifies the necessary steps for engaging in nuclear cooperation with another country.

- *Section 123a.* states that the proposed agreement is to include the terms, conditions, duration, nature, and scope of cooperation and lists nine criteria that the agreement must meet. It also contains provisions for the President to exempt an agreement from any of several criteria described in that section and includes details on the kinds of information the executive branch must provide to Congress.

- *Section 123b.* specifies the process for submitting the text of the agreement to Congress.
- *Section 123c.* specifies how Congress approves cooperation agreements that are limited in scope (e.g., do not transfer nuclear material or cover reactors larger than 5 MWe). This report does not discuss such agreements.
- *Section 123d.* specifies how Congress approves agreements that do cover significant nuclear cooperation (transfer of nuclear material or reactors larger than 5 MWe), including exempted agreements.

Section 123a., paragraphs (1) through (9), lists nine criteria that an agreement with a non-nuclear weapon state must meet unless the President determines an exemption is necessary. These include guarantees that

- safeguards on transferred nuclear material and equipment continue in perpetuity;
- full-scope International Atomic Energy Agency (IAEA) safeguards are applied in non-nuclear weapon states;
- nothing transferred is used for any nuclear explosive device or for any other military purpose; the United States has the right to demand the return of transferred nuclear materials and equipment, as well as any special nuclear material produced through their use, if the cooperating state detonates a nuclear explosive device or terminates or abrogates an IAEA safeguards agreement;
- there is no retransfer of material or classified data without U.S. consent;
- physical security on nuclear material is maintained;
- there is no enrichment or reprocessing by the recipient state of transferred nuclear material or nuclear material produced with materials or facilities transferred pursuant to the agreement without prior approval;
- storage for transferred plutonium and highly enriched uranium is approved in advance by the United States; and
- any material or facility produced or constructed through use of special nuclear technology transferred under the cooperation agreement is subject to all of the above requirements.

Exempted vs. Non-exempted Agreements

The President may exempt an agreement for cooperation from any of the requirements in Section 123a. if he determines that the requirement would be "seriously prejudicial to the achievement of U.S. non-proliferation objectives or otherwise jeopardize the common defense and security." The AEA provides different requirements, conditions, and procedures for exempt and non-exempt agreements.[4] To date, all of the Section 123 agreements in force are non-exempt agreements.[5] Prior to the adoption of P.L. 109-401, the Henry J. Hyde United States-India Peaceful Atomic Energy Cooperation Act of 2006, the President would have needed to exempt the nuclear cooperation agreement with India from some requirements of Section 123 a. P.L. 109-401, however, exempted nuclear cooperation with India from some of the AEA's requirements.[6]

Congressional Review

Under the AEA, Congress has the opportunity to review a 123 agreement for two time periods totaling 90 days of continuous session.[7] The President must submit the text of the proposed nuclear cooperation agreement, along with required supporting documents (including the unclassified NPAS) to the House Foreign Affairs Committee and the Senate Foreign Relations Committee. The President is to consult with the committees "for a period of not less than 30 days of continuous session." After this period of consultation, the President is to submit the agreement to Congress, along with the classified annex to the NPAS and a statement of his approval of the agreement and determination that it will not damage the national security interests of the United States. This action begins the second period, which consists of 60 days of continuous session. In practice, the President has sent the agreement to Congress at the beginning of the full 90-day period, which begins on the date of transmittal. Typically, the 60-day period has immediately followed the expiration of the 30-day period. The President transmits the text of the proposed agreement along with a letter of support with a national security determination, the unclassified NPAS, its classified annex, and letters of support for the agreement from the Secretary of State and the Nuclear Regulatory Commission.

If the President has not exempted the agreement from any requirements of Section 123a., it becomes effective at the end of the 60-day period unless, during that time, Congress adopts a joint resolution disapproving the agreement and the resolution becomes law. If the agreement is an exempted agreement, Congress must adopt a joint resolution of approval and it must become law by the end of the 60-day period or the agreement will not enter into force. At the beginning of this 60-day period, joint resolutions of approval or disapproval, as appropriate, are to be automatically introduced in each house. During this period, the committees are to hold hearings on the proposed agreement and "submit a report to their respective bodies recommending whether it should be approved or disapproved." If either committee has not reported the requisite joint resolution of approval or disapproval by the end of 45 days, it is automatically discharged from further consideration of the measure. After the joint resolution is reported or discharged, Congress is to consider it under expedited procedures, as established by Section 130.i. of the AEA.

Section 202 of P.L. 110-369, the United States-India Nuclear Cooperation Approval and Nonproliferation Enhancement Act, which President Bush signed into law October 8, 2008, amended Section 123 of the AEA to require the President to keep the Senate Foreign Relations Committee and the House Foreign Affairs Committee "fully and currently informed of any initiative or negotiations relating to a new or amended agreement for peaceful nuclear cooperation."

EXPORT LICENSING

The AEA sets out procedures for licensing exports to states with whom the United States has nuclear cooperation agreements. (Sections 126, 127, and 128 codified as amended at 42 U.S.C. 2155, 2156, 2157.) Each export of nuclear material, equipment, or technology requires a specific export license or other authorization. The Nuclear Regulatory Commission (NRC) is required to meet criteria in Sections 127 and 128 in authorizing export licenses. These criteria are as follows:

- Application of IAEA safeguards to any material or facilities proposed to be exported, material or facilities previously exported, and to any special nuclear material used in or produced through the use thereof (these are not full-scope safeguards, but safeguards required under Article III.2 of the nuclear Nonproliferation Treaty [NPT]).

- Nothing exported can be used for any nuclear explosive device or for research on or development of any nuclear explosive device.
- Recipient states must have adequate physical security on "such material or facilities proposed to be exported and to any special nuclear material used in or produced through the use thereof."
- Recipient states are not to retransfer exported nuclear materials, facilities, sensitive nuclear technology, or "special nuclear material produced through the use of such material" without prior U.S. approval.
- Recipient states may not reprocess or alter in form or content exported nuclear material or special nuclear material produced though the use of exported nuclear material without prior U.S. approval.
- The foregoing conditions must be applied to any nuclear material or equipment that is produced or constructed under the jurisdiction of the recipient by or through the use of any exported sensitive nuclear technology.
- Section 128 requires that recipient non-nuclear-weapon states must have full-scope IAEA safeguards.

The President must judge that the proposed export or exemption will "not be inimical to the common defense and security" or that any export of that type "would not be inimical to the common defense and security because it lacks significance for nuclear explosive purposes." The executive branch may also consider other factors, such as "whether the license or exemption will materially advance the nonproliferation policy of the United States by encouraging the recipient nation to adhere" to the NPT; whether "failure to issue the license or grant the exemption would otherwise be seriously prejudicial" to U.S. nonproliferation objectives; and whether the recipient nation has agreed to conditions identical to those laid out in Section 127.

Section 126b. (2) contains a provision for the President to authorize an export in the event that the NRC deems that the export would not meet Section 127 and 128 criteria. The President must determine "that failure to approve an export would be seriously prejudicial to the achievement of U.S. nonproliferation objectives or otherwise jeopardize the common defense and security." In that case, the President would submit his executive order, along with a detailed assessment and other documentation, to Congress for 60 days of continuous session. After 60 days of continuous session, the export would go through unless Congress were to adopt a concurrent resolution of disapproval.[8]

Section 128b.(2) contains a provision for the President to waive termination of exports by notifying the Congress that the state has adopted full-scope safeguards or that the state has made significant progress toward adopting such safeguards, or that U.S. foreign policy interests dictate reconsideration. Such a determination would become effective unless Congress were to adopt a concurrent resolution of disapproval within 60 days of continuous session.

Additionally, Section 129b.(1) forbids the export of "nuclear materials and equipment or sensitive nuclear technology" to any country designated as a state sponsor of terrorism.[9] Section 129b.(3) allows the President to waive this provision.

SUBSEQUENT ARRANGEMENTS

Section 131 of the Atomic Energy Act details procedures for subsequent arrangements to nuclear cooperation agreements concluded pursuant to Section 123. Such arrangements are required for forms of nuclear cooperation requiring additional congressional approval, such as transfers of nuclear material or technology and recipient states' enrichment or reprocessing of nuclear materials transferred pursuant to the agreement. Subsequent arrangements may also include arrangements for physical security, storage, or disposition of spent nuclear fuel; the application of safeguards on nuclear materials or equipment; or "any other arrangement which the President finds to be important from the standpoint of preventing proliferation."

Before entering into a subsequent arrangement, the Secretary of Energy must publish in the *Federal Register* a determination that the arrangement "will not be inimical to the common defense and security." A proposed subsequent arrangement shall not take effect before 15 days after publication of both this determination and notice of the proposed arrangement. The Secretary of State is required to prepare an unclassified Nuclear Proliferation Assessment Statement (NPAS) if, "in the view of" the Secretary of State, Secretary of Energy, Secretary of Defense, or the Nuclear Regulatory Commission, a proposed subsequent arrangement "might significantly contribute to proliferation." The Secretary of State is to submit the NPAS to the Secretary of Energy within 60 days of receiving a copy of the proposed subsequent arrangement. The President may waive the 60-day requirement if the Secretary of State so requests, but must notify both the House Foreign Affairs Committee and Senate Foreign Relations Committee of any such waiver and the justification for it. The Secretary of Energy may not enter into the subsequent arrangement before receiving the NPAS.

Section 131 specifies requirements for certain types of subsequent arrangements. Section 131b. describes procedures for the executive branch to follow before entering into a subsequent arrangement involving the reprocessing of U.S.-origin nuclear material or nuclear material produced with U.S.-supplied nuclear technology. These procedures also cover subsequent arrangements allowing the retransfer of such material to a "third country for reprocessing" or "the subsequent retransfer" of more than 500 grams of any plutonium produced by reprocessing such material. The Secretary of Energy must provide both the House Foreign Affairs Committee and Senate Foreign Relations Committee with a report describing the reasons for entering into the arrangement. Additionally, 15 days of continuous session must elapse before the Secretary may enter into the arrangement, unless the President judges that "an emergency exists due to unforeseen circumstances requiring immediate entry" into the arrangement. In such a case, the waiting period would be 15 calendar days.

If a subsequent arrangement described in the above paragraph involves a facility that has not processed spent nuclear reactor fuel prior to March 10, 1978 (when the Nuclear Nonproliferation Act of 1978 was enacted), the Secretaries of State and Energy must judge that the arrangement "will not result in a significant increase of the risk of proliferation." In making this judgment, the Secretaries are to give "foremost consideration ... to whether or not the reprocessing or retransfer will take place under conditions that will ensure timely warning to the United States of any diversion well in advance of the time at which the non-nuclear-weapon state could transform the diverted material into a nuclear explosive device."[10] For a subsequent arrangement involving reprocessing in a facility that has processed spent nuclear reactor fuel prior to March 10, 1978, the Secretary of Energy will "attempt to ensure" that reprocessing "shall take place under conditions" that would satisfy the timely-warning conditions described above. Section 131f. specifies procedures for congressional approval of subsequent arrangements involving the storage or disposition of foreign spent nuclear fuel in the United States.

Section 133 states that, before approving a subsequent arrangement involving certain transfers of special nuclear material, the Secretary of Energy must consult with the Secretary of Defense "on whether the physical protection of that material during the export or transfer will be adequate to deter theft, sabotage, and other acts of international terrorism which would result in the diversion of that material."[11] If the Secretary of Defense determines that "the export or transfer might be subject to a genuine terrorist threat," that Secretary is required to provide a written risk assessment of the risk and a "description of the actions" that he or she "considers necessary to upgrade physical protection measures."

Examples of Subsequent Arrangements

U.S.-Japan Agreement

The first test of the subsequent arrangement provisions came in August 1978, when the Department of Energy informed the House and Senate foreign relations committees of a Japanese request for approval of the transfer of spent fuel assemblies from Japan to the United Kingdom for reprocessing. This was the first "subsequent arrangement" approved. The United States and Japan entered into similar arrangements until 1988, when the two governments revised their nuclear cooperation agreement. That agreement included an "implementing agreement," which provided 30-year advance consent for the transfer of spent fuel from Japan to Europe for reprocessing. While controversial, Congress did not block the nuclear cooperation agreement.

A subsequent arrangement was also necessary for the sea transport from Europe to Japan of plutonium that had been separated from the Japanese spent fuel. The Department of Energy approved a Japanese request for 30-year advance consent for the sea transport of plutonium. It was submitted to Congress as a subsequent arrangement, and took effect in October 1988.

U.S.-India Agreement

The U.S. nuclear cooperation agreement with India grants New Delhi consent to reprocess nuclear material transferred pursuant to the agreement, as well as "nuclear material and byproduct material used in or produced through the use of nuclear material, non-nuclear material, or equipment so transferred." However, the agreement also includes a requirement that India first build a new national reprocessing facility to be operated under IAEA safeguards. The two countries signed a subsequent arrangement July 30, 2010, which governs the procedures for operating two new reprocessing facilities in India. The agreement also describes procedures for U.S. officials to inspect and receive information about physical protection measures at the new facilities. The arrangement would not have taken effect if Congress had adopted a joint resolution of disapproval within 30 days of continuous session; Congress did not adopt such a resolution.[12] If India were to construct any additional facilities to reprocess fuel from U.S.-supplied reactors, a new subsequent arrangement would need to be submitted to Congress.

Iran-Related Restrictions

The Comprehensive Iran Sanctions, Accountability, and Divestment Act (CISADA) of 2010 (P.L. 111-195), which became law July 1, 2010, contains additional restrictions on licensing nuclear exports to countries with entities that have been sanctioned for conducting certain types of energy-related transactions with Iran. Section 102 a.(2)(A) of the law states that "no license may be issued for the export, and no approval may be given for the transfer or retransfer" of "any nuclear material, facilities, components, or other goods, services, or technology that are or would be subject to an agreement for cooperation between the United States" and such countries. Section 102 a.(2)(B), however, allows the President to waive these restrictions. Section 102 a.(2)(C), allows the President to authorize licenses for nuclear exports "on a case-by-case basis" to entities (which have not been sanctioned) in countries subject to the restrictions described above.[13]

TERMINATION OF COOPERATION

Section 129a. of the AEA requires that the United States end exports of nuclear materials and equipment or sensitive nuclear technology to any non-nuclear-weapon state that, after March 10, 1978, the President determines to have detonated a nuclear explosive device; terminated or abrogated IAEA safeguards; materially violated an IAEA safeguards agreement; or engaged in activities involving source or special nuclear material and having "direct significance" for the manufacture or acquisition of nuclear explosive devices, and "has failed to take steps which, in the President's judgment, represent sufficient progress toward terminating such activities."

Section 129a. also requires that the United States halt exports to any nation the President determines to have materially violated the terms of an agreement for cooperation with the United States; assisted, encouraged, or induced any non-nuclear weapon state to obtain nuclear explosives or the materials and technologies needed to manufacture them; or re-transferred or entered into an agreement for exporting reprocessing equipment, materials, or technology to a non-nuclear weapon state, unless in connection with an international agreement to which the United States subscribes.

The President can waive termination of exports if he determines that "cessation of such exports would be seriously prejudicial to the achievement of United States nonproliferation objectives or otherwise jeopardize the common defense and security." The President must submit his determination to Congress, which is then referred to the House Committee on Foreign Affairs and the Senate Foreign Relations Committee for 60 days of continuous session. The determination becomes effective unless Congress adopts a joint resolution opposing the determination.

LEGISLATION IN THE 112TH CONGRESS

Congress has intensely debated the requirements of the Atomic Energy Act for nuclear cooperation agreements, most notably prior to the approval of the 2008 U.S.-India nuclear cooperation agreement.

During the past two years, several hearings have been held, particularly by the House Foreign Affairs Committee and its Nonproliferation, Terrorism, and Trade Subcommittee, to discuss whether Congress should legislate stronger nonproliferation standards and congressional oversight for nuclear cooperation agreements. This issue continues to be a subject of attention in the 112th Congress.

S. 109

On January 25, 2011, Senator John Ensign introduced S. 109, which was discharged from the Committee on Finance and forwarded to the Committee on Foreign Relations on February 3. This bill would change current congressional review requirements by mandating positive congressional approval of all Section 123 agreements. It would also require a joint resolution of approval for subsequent arrangements (under Section 131 a(1)), and a nuclear proliferation assessment statement to be submitted along with proposals for subsequent arrangements.

S. 109 would also seek to strengthen congressional oversight of the negotiation of nuclear cooperation agreements.

The bill would require the President to keep the House Foreign Affairs Committee and the Senate Foreign Relations Committee "fully and currently informed of any initiative or negotiations" relating to a new or amended nuclear cooperation agreement prior to any public presidential announcements. S. 109 would also require the President to consult with relevant committees "not less than 15 calendar days" after the start of negotiations, and on a monthly basis during negotiations. As noted, current law requires only that the President keep Congress "fully and currently informed of any initiative or negotiations relating to" a new nuclear cooperation agreement.

H.R. 1280

On March 31, 2011, the chairman of the House Foreign Affairs Committee, Representative Ileana Ros-Lehtinen, introduced H.R. 1280. The bill, which has bipartisan co-sponsorship, would increase congressional oversight over the negotiation and approval of 123 agreements and subsequent arrangements.

For example, it would change the Atomic Energy Act's congressional review requirements by requiring that Congress adopt a joint resolution of approval within 60 days of continuous session before any nuclear cooperation agreement could go into effect (the 60 days would be part of the 90-day period of continuous session described above). The bill would also require a joint resolution of approval for subsequent arrangements under nuclear cooperation agreements. Additionally, H.R. 1280 contains S. 109's notification and consultation provisions.

H.R. 1280 would also add new nonproliferation criteria to Section 123a.[14] Section 1 of the bill would require governments wishing to conclude nuclear cooperation agreements with the United States to

- guarantee that third-country nationals will not be permitted, without prior U.S. consent, access to facilities, material or technology transferred under the agreement;
- forswear "enrichment or reprocessing activities, or acquisition or construction of facilities for such activities," if the country did not operate enrichment or reprocessing facilities as of April 1, 2011;
- accede to and fully implement the Chemical Weapons Convention, the Biological Weapons Convention, and "all other international agreements to which the United States is a party related to export controls on nuclear, chemical,
- biological, and advanced conventional weapons";

- have an effective export control system and implement UN Security Council Resolution 1540[15];
- be fully compliant with United Nations conventions and Security Council resolutions related to nonproliferation, including the Convention of the Physical Protection of Nuclear Material and the Nuclear Terrorism Convention;
- not be a Destination of Diversion Concern as defined by CISADA (P.L. 111- 195)[16];
- closely cooperate with the United States to prevent states sponsors of terrorism from acquiring nuclear, biological, chemical, or advanced conventional weapons;
- have an Additional Protocol to its safeguards agreement with the IAEA fully in force;
- agree to refrain from using nuclear material or equipment acquired from nonU.S. sources (in addition to such material or equipment obtained from the United States) for a "nuclear explosive device" or any other military purpose; and
- guarantee that it will maintain "adequate physical security" on all special nuclear material, in addition to such material produced using material or equipment obtained from the United States.

The bill contains several other significant provisions. Section 2 would prohibit any U.S. assistance, other than humanitarian aid, to a cooperating country that has withdrawn from the NPT. Moreover, the country would be asked to return any material, equipment, or components transferred under a nuclear cooperation agreement (on or after the bill's enactment date), and any special fissionable material produced using material or equipment previously transferred. Section 7 of the bill would prohibit the President from issuing an export license pursuant to a nuclear cooperation agreement unless the cooperating country had liability protection for United States nuclear suppliers "equivalent to the liability protection specified under the Convention on Supplementary Compensation for Nuclear Damage." Section 3 of the bill would require the executive branch to produce a report detailing the nonproliferation conditions of other nuclear suppliers. Section 5 would require termination of nuclear exports to a cooperating country that had been "determined to be a 'country of proliferation concern'" pursuant to the 2010 National Defense Authorization Act.[17]

H.R. 1280, Amended

On April 14, the House Foreign Affairs Committee amended H.R. 1280. An agreement lacking a legally binding requirement for the cooperating country to refrain from enrichment or reprocessing activities (or constructing facilities for such activities) could not enter into effect, according to the amended bill, unless Congress were to adopt a joint resolution of approval within 60 days of continuous session. An agreement including such a provision would become effective after 60 days of continuous session unless Congress were to adopt a joint resolution of disapproval (the 60 days referred to in this and the previous sentence would be part of the 90-day period of continuous session described above).[18]

The amended version of H.R. 1280 would also prohibit any U.S. assistance, other than humanitarian aid, to a cooperating country that has withdrawn from the NPT. However, only a cooperating country withdrawing after enactment of the proposed law would be subject to this prohibition. The amended bill also contains H.R. 1320's provisions regarding U.S. assistance as they pertain to state sponsors of proliferation and states without Additional Protocols in force (see below).

The committee also adopted two amendments sponsored by Representative Brad Sherman. The first would amend Section 123 of the Atomic Energy Act to require that a cooperating country have a commitment to maintain or enact, as circumstances dictate, a "legal regime providing for adequate protection from civil liability that will allow for the participation of United States suppliers in any effort by the country to develop civilian nuclear power." The country would be required to have such a regime in place no later than one year after a nuclear cooperation agreement's entry into force.

The second expresses the sense of Congress that the President

> should ensure that participation in international nuclear programs conducted by the United States is limited to the greatest extent practicable to governmental and nongovernmental participants from countries that have adopted nonproliferation provisions in their nuclear cooperation and nuclear export control policies comparable to the policies specified in section 123 of the Atomic Energy Act (42 U.S.C. 2153), as amended by this Act.

The amended H.R. 1280 and H.Rept. 112-507 were reported out of committee on May 30, 2012. H.R. 1280 was referred to the House Rules

Committee and the House Committee on Energy and Commerce on May 30, 2012. Those committees have until October 1, 2012, to consider the bill.

The Administration's Response

The State Department articulated its opposition to H.R. 1280 in a July 15, 2011, fact sheet which argued that, were it to become law, the bill would damage nuclear cooperation agreements' "effectiveness as a nonproliferation tool."[19] According to the State Department, it "would be difficult" for partners to nuclear cooperation agreements to accept the new conditions described in the bill. Moreover, "the United States would not be able to accept some of these conditions on a reciprocal basis, as would be required by virtually all of our partners," the fact sheet says, adding that the new requirements "would likely cause" such countries "to consider nuclear cooperation with other countries instead of the United States, thus preventing the achievement of the bill's nonproliferation objectives."

The State Department also described H.R. 1280's provision prohibiting third-party nationals' access to U.S.-supplied nuclear reactors, equipment, or technology without prior U.S. consent as "impractical," arguing that it would assert a U.S. "right to 'approve'" IAEA inspectors from third countries to gain access to such reactors, equipment, and technology, thereby interfering with the IAEA's ability to apply safeguards in the recipient country.

The fact sheet also criticizes H.R. 1280's liability provisions, arguing that they "would create a legal obligation to meet a vague standard, to be judged by others, that few states would be willing to accept," adding that "decisions about what constitutes adequate liability protection are commercial determinations that are best made by U.S. suppliers in the context of a particular transaction."

H.R. 1320

House Foreign Affairs Committee Ranking Member Howard Berman introduced H.R. 1320, cosponsored by Representative Brad Sherman, on April 1, 2011. The bill is divided into two sections—strengthening nonproliferation activities (Title I) and strengthening nuclear energy cooperation and nonproliferation (Title II).

Title I contains several provisions regarding U.S. assistance to other countries. Section 103 would require the U.S. government, before providing assistance, to consider whether a country has an Additional Protocol in force.

Section 104 would prohibit U.S. assistance to state sponsors of proliferation of weapons of mass destruction.[20] Section 101 contains H.R. 1280's provisions regarding countries that have withdrawn from the NPT.

H.R. 1320 would also require payment of U.S. annual dues to the IAEA's regular budget, as well as a report detailing nonproliferation conditions of foreign nuclear suppliers. It would also require periodic security inspections on U.S. exports of nuclear material "to ensure that adequate physical safeguards and accounting measures are in effect for such nuclear material."

Title II would add conditions for new peaceful nuclear cooperation agreements. The bill contains H.R. 1280's provisions regarding additional protocols; third-country nationals' access to nuclear facilities, material, or technology; and the requirement that potential cooperating countries agree to refrain from using nuclear material or equipment acquired from non-U.S. sources for military purposes. Title II also contains a variation of H.R. 1280's enrichment and reprocessing provisions, but H.R. 1320 would not require a potential cooperating country to forswear these activities if they would be "pursued as part of a multilateral consortium of countries" that are NPT state-parties in good standing. The bill would require the consortium to have "the explicit support of the United States." Unlike H.R. 1280, H.R. 1320 does not specify that these requirements apply only to countries possessing enrichment and reprocessing facilities as of April 1, 2011. H.R. 1320 would also require the President to determine that a potential cooperating country had not engaged in any significant proliferation activities and was cooperating with the United States to halt proliferation.

The bill would also amend congressional review requirements. An agreement that complied with all of the nonproliferation criteria could be brought into effect 60 legislative days after the President had submitted the agreement to Congress, unless Congress were to adopt a joint resolution during those 60 days stating that Congress did not favor the agreement. If an agreement did not comply with the nonproliferation criteria and/or the required presidential determinations could not be made, Congress would be required to adopt a joint resolution of approval before the agreement entered into effect. Renewal agreements meeting the nonproliferation criteria and presidential determinations would need to sit before Congress for 30 calendar days before they could be brought into effect. A 45-legislative-day congressional review would be required for subsequent arrangements, after which time the arrangement would come into effect if Congress had not enacted a joint resolution of disapproval.

As with H.R. 1280, the bill says the President could not issue a license for export under the agreement to a cooperating country without liability protection for U.S. nuclear suppliers "equivalent to the liability protection specified under the Convention on Supplementary Compensation for Nuclear Damage."

H.R. 1320 would also give the President the option to make loan guarantees to countries that forego enrichment or reprocessing and with whom the United States has a 123 agreement.

FY2012 Intelligence Authorization Act (P.L. 112-87)

The Senate version of the FY2012 Intelligence Authorization Act (S. 1458) contained amendments to Section 123 of the AEA (as amended). The bill was reported out of committee in August 2011. The proposed amendments found in Section 305 of the bill would update language to reflect the establishment of the Director of National Intelligence and require additional input by the intelligence community in developing the NPAS for a 123 agreement. The Senate report explains:

> Section 305 requires the Secretary of State and the DNI to provide an unclassified NPAS to the President. The first two sections of the NPAS, which mirror the current requirements, shall be prepared by the Secretary of State, in consultation with the Director of National Intelligence. A newly-required third section shall be prepared by the DNI, in consultation with the Secretary of State, and provide a comprehensive analysis of the country's export control system with respect to nuclear-related matters. A classified annex shall accompany the NPAS. The NPAS and its classified annex shall be provided to the congressional intelligence committees as well as the congressional foreign relations committees.

The companion House bill (H.R. 1892) was passed on September 9, 2011. H.R. 1892 would amend Section 102A of the National Security Act of 1947 by adding the DNI's role in producing the NPAS, as such:

> Nuclear Proliferation Assessment Statements Intelligence Community Addendum- The Director of National Intelligence, in consultation with the heads of the appropriate elements of the intelligence community and the Secretary of State, shall provide to the President, the congressional intelligence committees, the Committee on Foreign Affairs of the House of

> Representatives, and the Committee on Foreign Relations of the Senate an addendum to each Nuclear Proliferation Assessment Statement accompanying a civilian nuclear cooperation agreement, containing a comprehensive analysis of the country's export control system with respect to nuclear-related matters, including interactions with other countries of proliferation concern and the actual or suspected nuclear, dual-use, or missile-related transfers to such countries.

The Intelligence Authorization Act for Fiscal Year 2012 (P.L. 112-87), Section 305, "Preparation of Nuclear Proliferation Assessment Statements," included the H.R. 1892 language on Nuclear Proliferation Assessment Statements cited above.

H.R. 1905 and H.R. 2105

These two bills, passed by the House on December 14, 2011, would restrict nuclear cooperation with countries sanctioned for providing certain types of technology to Iran, Syria, or North Korea. Section 105 (b)(2) of H.R. 1905 would prohibit export licenses and transfers of "any nuclear material, facilities, components, or other goods, services, or technology that are or would be subject to" a nuclear cooperation agreement between the United States and a government which has "primary jurisdiction" over a person subject to sanctions (described elsewhere in the bill) for "an activity" relating to "the acquisition or development of nuclear weapons or related technology or of missiles or advanced conventional weapons that are designed or modified to deliver a nuclear weapon." These sanctions, however, shall not apply to such a government if the President "determines and notifies the appropriate congressional committees" that the government "does not know or have reason to know about the activity" or "has taken, or is taking, all reasonable steps necessary to prevent a recurrence of the activity and to penalize the person for the activity." The bill also includes a provision which would allow transfers described above on a "case-bycase basis."

Section six of H.R. 2105 would prohibit the executive branch from submitting to Congress a nuclear cooperation agreement with any country "that is assisting the nuclear program of Iran, North Korea, or Syria, or transferring advanced conventional weapons or missiles" to those countries. The bill would also prohibit a nuclear cooperation agreement with such an assisting country from entering into force, as well as prohibit export licenses and transfers of "any nuclear material, facilities, components, or other goods, services, or technology that would be subject to" a nuclear cooperation agreement. All of these forms of nuclear cooperation would be prohibited until the President issued a determination and a report to Congress that Iran, North Korea, or Syria (as the case may be) "has ceased its efforts to design, develop, or acquire a nuclear explosive device or related materials or technology" or the assisting government has suspended the offending activities and "committed to maintaining that suspension" until Iran, North Korea, or Syria (as the case may be) "has implemented measures that would permit the President to make the determination" regarding those countries' weapons programs. The bill names Russia as a country that is "assisting the nuclear program of Iran, North Korea, or Syria or transferring advanced conventional weapons or missiles" to those countries.

S. 1048

The Senate passed H.R. 1905 on May 21, 2011, substituting Section 205(c) of S. 1048, which was introduced May 23, 2011, and about which the Committee on Banking, Housing, and Urban Affairs held hearings October 13, would require the Secretary of Energy, the Secretary of Commerce, and the Nuclear Regulatory Commission to certify to Congress that issuing an export license pursuant to a nuclear cooperation agreement would not "permit the transfer" of "any good or technology that the President determines is used, or is likely to be used, for military applications" to Iran, North Korea, or Syria.

As of mid-June 2012, the Senate and House versions of H.R. 1905 were awaiting conference action.

End Notes

1 The Atomic Energy Act (AEA) was amended by the Nuclear Nonproliferation Act of 1978 (NNPA) (P.L. 95-242) to include stringent nonproliferation requirements for significant U.S. nuclear exports. For example, the act required nonnuclear-weapon states to have full-scope International Atomic Energy Agency safeguards as a condition for entering into nuclear cooperation agreements with the United States. For existing and future agreements, the NNPA added a provision for Congress to review export licenses. The act also included a provision for halting exports if a country tested a nuclear device, violated safeguards agreements, or continued nuclear weapons-related activities.

2 Section 57 b. (2) of the AEA allows for limited forms of nuclear cooperation related to the "development or production of any special nuclear material outside of the United States"

without a nuclear cooperation agreement if that activity has been authorized by the Secretary of Energy following a determination that it "will not be inimical to the interest of the United States." Agreements governing such cooperation are also known as "Section 810" agreements, after 10 Code of Federal Regulations Part 810. A nuclear cooperation agreement is not required for transmission of nuclear-related information, except for restricted data. The term "restricted data," as well as other terms used in the statute, is defined in 42 U.S.C. §2014. "Restricted data" means "all data concerning (1) design, manufacture, or utilization of atomic weapons; (2) the production of special nuclear material; or (3) the use of special nuclear material in the production of energy." Restricted data, however does not "include data declassified or removed from the Restricted Data [sic] category" pursuant to the AEA. A nuclear cooperation agreement is necessary, though not necessarily sufficient, to permit the transfer of restricted data.

[3] "Special nuclear material" means (1) plutonium, uranium enriched in the isotopes 233 or 235, and any other material that is determined to be special nuclear material, but does not include source material, or (2) any material artificially enriched by any of the foregoing, but does not include source material.

[4] Nuclear cooperation agreements with nuclear weapon states recognized by the NPT are provided for in the AEA, and are therefore non-exempt agreements. The NPT defines nuclear weapon states as those that exploded a nuclear weapon or other nuclear explosive device prior to January 1, 1967: China, France, Russia, the United Kingdom, and the United States.

[5] The United States has concluded more than 20 bilateral nuclear cooperation agreements, as well as similar agreements with the European Atomic Energy Community and the IAEA. See Nuclear Commerce: Governmentwide Strategy Could Help Increase Commercial Benefits from U.S. Nuclear Cooperation Agreements with Other Countries, Government Accountability Office, GAO-11-36, November 2010.

[6] See CRS Report RL33016, U.S. Nuclear Cooperation with India: Issues for Congress, by Paul K. Kerr. The agreement entered into force in December 2008.

[7] When calculating periods of "continuous session" under the AEA, every calendar day is counted, including Saturdays and Sundays. Only days on which either chamber has adjourned for more than three days pursuant to the adoption a concurrent resolution authorizing the adjournment do not count toward the total. If Congress adjourns its final session sine die, continuity of session is broken, and the count must start anew when it reconvenes.

[8] In light of the Supreme Court's 1983 decision in INS v. Chadha, passing a concurrent resolution could invite a legal challenge because it is arguably unconstitutional. Although not provided for in the AEA, Congress could choose to pass a joint resolution of disapproval or a bill stating in substance it did not approve.

[9] Section 129b. (2) states that the prohibitions described in the previous section "shall not apply to exports, reexports, transfers, or retransfers of radiation monitoring technologies, surveillance equipment, seals, cameras, tamper-indication devices, nuclear detectors, monitoring systems, or equipment necessary to safely store, transport, or remove hazardous materials ... except to the extent that such technologies, equipment, seals, cameras, devices, detectors, or systems are available for use in the design or construction of nuclear reactors or nuclear weapons."

[10] These provisions also applies to facilities that, prior to March 18, 1978, did not have a subsequent arrangement for reprocessing.

[11] This section applies to "the export or transfer of more than 2 kilograms of plutonium or more than 5 kilograms of uranium enriched to more than 20 percent in the isotope 233 or the isotope 235."

[12] Section 201 of the United States-India Nuclear Cooperation Approval and Nonproliferation Enhancement Act (P.L. 110-369), which approved the U.S.-India cooperation agreement, specifies procedures (different from those described in the Atomic Energy Act) for Congress to consider subsequent arrangements to that agreement.

[13] For details on these sanctions, see CRS Report RS20871, Iran Sanctions, by Kenneth Katzman.

[14] This summary is not meant to be exhaustive, and the full bill text should be consulted.

[15] That resolution requires states to "take and enforce effective measures to establish domestic controls to prevent the proliferation of nuclear, chemical, or biological weapons and their means of delivery, including by establishing appropriate controls over related materials."

[16] CISADA states that the President "shall designate a country as a Destination of Diversion Concern if the President determines that the government of the country allows substantial diversion of goods, services, or technologies" described in the law "through the country to Iranian end-users or Iranian intermediaries."

[17] That law defines a country of proliferation concern as "any country identified by the Director of Central Intelligence as having engaged in the acquisition of dual-use and other technology useful for the development or production of weapons of mass destruction (including nuclear weapons, chemical weapons, and biological weapons) or advanced conventional munitions."

[18] The 60-day period would apply to new agreements. For a renewal agreement, Congress would need to adopt a joint resolution of disapproval within 30 days of continuous session. The amended text defines a "renewal agreement" as a nuclear cooperation agreement with a country which has previously entered into such an agreement with the United States. A "new" nuclear cooperation agreement, according to the draft legislation, is one with a country which has not previously entered into such an agreement with the United States.

[19] http://www.state.gov/t/isn/rls/fs/168659.htm.

[20] The bill would require the Secretary of State to determine "that the government of the country has repeatedly provided support for acts of proliferation of equipment, technology, or materials to support the design, acquisition, manufacture, or use of weapons of mass destruction."

INDEX

A

B

C

J

K

L

M

N

O

T

U

V

W

Y